EU ENVIRONMENTAL POLICIES IN SUBNATIONAL REGIONS

EU Environmental Policies in Subnational Regions

The case of Scotland and Bavaria

ANTJE C. K. BROWN
University of Aberdeen, Scotland

Ashgate

Aldershot • Burlington USA • Singapore • Sydney

Published by
Ashgate Publishing Limited
Gower House
Croft Road
Aldershot
Hampshire GU11 3HR
England

Ashgate Publishing Company
131 Mainstreet
Burlington, VT 05401-5600 USA

Ashgate website: http://www.ashgate.com

British Library Cataloging in Publication Data
Brown, Antje C. K.
    EU environmental policies in subnational regions : the case
    of Scotland and Bavaria. - (Ashgate studies in
    environmental policy and practice)
    1.European Union 2. Environmental policy - European Union
    countries 3.Environmental protection - Scotland
    4.Environmental protection - Germany - Bavaria
    I.Title
    363.7'0094

Library of Congress Cataloging in Publication Data
Brown, Antje C.K., 1967-
    EU environmental policies in subnational regions : the case of Scotland and Bavaria /
Antje C.K. Brown.
        p. cm. -- (Ashgate studies in environmental policy and practice)
    Includes bibliographical references and index.
    ISBN 0-7546-1734-3
        1. Environmental policy--European Union countries. 2. Environmental
policy--Scotland--Case studies. 3. Environmental policy--Germany--Bavaria--Case
studies. I. Title. II. Series.

    GE190.E85 B76 2002
    363.7'056'094--dc21   2001053714

ISBN 0 7546 1734 3

Printed and bound by Athenaeum Press, Ltd.,
Gateshead, Tyne & Wear.

Contents

List of Figures and Tables

Acknowledgements

I would like to thank my former doctoral supervisor Elizabeth Bomberg for her invaluable advice, guidance and encouragement in completing this research. I am indebted to Stephen Ingle and the staff of the Politics Department, University of Stirling, who supported me immensely throughout my research. I am also grateful to my examiners Susan Baker and Eric Shaw who provided me with most helpful comments. I would also like to thank the Carnegie Trust, the Politics Department in Stirling and the Department of Politics and International Relations, University of Aberdeen, for their support and funding of research visits without which I could not have completed my research. My thanks go also to the editors of Ashgate Publishers who have given me the opportunity to publish my doctoral thesis. Needless to say I am most grateful to the numerous interviewees in Brussels, Scotland and Bavaria who have been very patient and helpful with my enquiries. Their contributions are very much appreciated. Of course, any errors that still remain should not be associated with any of the names just mentioned. They are entirely my own.

Finally, I would like to thank my husband Gerry and my family in Berlin for their amazing support and interest in my research.

Antje C. K. Brown

November 2001

List of Abbreviations

AT	Treaty of Amsterdam
BATNEEC	Best available technology not entailing excessive costs
BImschG	Bundes-Immissionsschutz Gesetz*
BMU	Bundesministerium für Umwelt, Naturschutz und Reaktorsicherheit
Bund	Bund Naturschutz in Bayern e.V.
BVerwG	Bundesverwaltungs-Gericht*
CDU	Christlich Demokratische Union Deutschlands
CEBIS	Centre for Environment & Business in Scotland
CFI	Court of First Instance
CFSP	Common Foreign and Security Policy
CO	Carbon Monoxide
CO2	Carbon Dioxide
CoR	Committee of the Regions

COREPER	Committee of Permanent Representatives in the EU
CoSLA	Convention of Scottish Local Authorities
CSU	Christlich Soziale Union in Bayern
DG	Commission Directorate-Generale
DG VII	Transport
DG XI	Environment, Nuclear Safety and Civil Protection
DGXVI	Regional Policy
DoE	UK Department of the Environment (since 1997: DETR Department of the Environment, Transport and the Regions)
EAP	Environmental Action Programme
EC	European Community
ECJ	European Court of Justice
EcoSoc	Economic and Social Committee
ECSC	European Coal and Steel Community
EEA	European Environmental Agengy
EEC	European Economic Community

EIA	Environmental Impact Assessment
EMU	Economic and Monetary Union
EP	European Parliament
ERM	Exchange Rate Mechanism
ES	Environmental Statement
ESEC	East of Scotland European Consortium
EU	European Union
FDP	Freie Demokratische Partei Deutschlands
FoEScotland	Friends of the Earth Scotland
FPTP-system	First past the post electoral system
FRG	Federal Republic of Germany
G7	Group of Seven
GDP	Gross Domestic Product
GDR	German Democratic Republic
HMIPI	Her Majesty's Industrial Pollution Inspectorate
ICM	Integrated Catchment Management

IMPEL	EU Network for the Implementation and Enforcement of Environmental Law
IPC	integrated pollution control
JHA	Justice and Home Affairs
LBV	Landesverban für Vogelschutz in Bayern e.V.
LEEP	Lothian and Edinburgh Environmental Partnership
LfU	Bayerisches Landesamt für Umweltshutz
MEP	Member of European Parliament
MP	Member of Parliament
MSP	Member of Scottish Parliament
NGO	Non-governmental Organisation
NO2	Nitrogen Dioxide
NSMs	new social movements
O.J.	Official Journal of the European Communities
PPP	Polluter Pays Principle
PR-system	Proportional representation electoral system

REEFs	Regional Environmental Education Forums
RPBs	River Purification Boards
RT	Rome Treaty
SEA	Single European Act
SEA	Strategic Environmental Assessment
SEEC	Scottish Environmental Education Council
SEPA	Scottish Environmental Protection Agency
SI	Statutory Instrument
SMEs	Small and Medium-sized Enterprises
SNP	Scottish Nationalist Party
SNA	Sub National Authority
SNH	Scottish Natural Heritage
SO2	Sulphur Dioxide
SPD	Sozialdemokratische Partei Deutschlands
StMLU	Bayerisches Staatsministerium für Landesentwicklung und Umweltfragen*

TEU	Treaty on European Union (or Maastricht Treaty)
UK	United Kingdom
UKREP	United Kingdom Permanent Representation in the EU
UN	United Nations
UPVG	Umweltverträglichkeits-prüfungsgesetz*
USA	United States of America
VerwVerfG	Verwaltungsverfahrens-Gesetz*

(*) = See 'Glossary of German Terms' for English translation

Glossary of German Terms

Amtsblätter: gazette

Angst: anxiety

Bayerisches Staatsministerium für Landesentwicklung und Umweltfragen: Bavarian Ministry for Planning, Development and Environmental Matters.

Beobachter der Länder: observer for the Länder in the EU

Beschleunigung von Genehmigungsverfahren: the speeding up of planning procedures

Bezirkstag: district assembly

Biokartierung: identification and regular assessment of natural habitats

Bundes-Immissionsschutz-Gesetz: Federal Law for the Regulation of Large Combustion Plants.

Bundesrat: upper (regional) chamber

Bundestag: lower (Federal) chamber

Bundesverwaltungsgericht: Federal Administrative Court

Eigenverantwortung: self-reliance, autonomy

Ernüchterung: viewing developments with certain soberness/ disillusionment

federführend: in charge

Flächenstaat: state with large natural, farming and forestry
                            resources

Freistaat Bayern: Free State of Bavaria

Gemeinderat: Community Council

Gemeinschaftsfreundlich: community/ Euro-friendly

Gesetz: Act of Parliament

Gesetzgeber: 'the legislator', legislature

Gewerbeordnung: trade ordinance

Grüne Punkt, Der: German recycling scheme of waste packaging

Grünen, Die: German Green Party (since merger with East
                            German sister party *Die Grünen/ Bündnis 90*)

jederman: anybody

konkrete Möglichkeit: definite possibility

Kreisfreie Stadt: town

Länder: Federal States

Landkreis: county

Landtag: state parliament

Ministerpräsident: head of state government

Musterverwaltung: 'example' or 'perfect' administration

nachvollziehen: to comprehend information

Neue Soziale Bewegungen: new social movements

Neutralitätsverständnis: sense of neutrality

Politikverflechtung: political inter-connectedness

Popularklagerecht: legal right for the general public to challenge decisions.

Rechtsverordnung: regulation

Regierungsbezirk: district

Senat: Bavarian Senate of Elders

Spitzenverbände: umbrella organisations/ associations

Stand der Technik: state of the art

stiefmütterlich: treating others like a stepmother treats a stepchild.

Überreguliertheit: over-regulation

Umwelt: (natural) environment

Umweltpakt: environmental pact between the Bavarian State Government and representatives of the private sector in Bavaria

Umweltverträglichkeitsprüfung: assessing the natural environment's ability to absorb projects and their damaging impacts

Verwaltungs-Verfahrens-Gesetz: Federal Law regulating Administrative Procedures

Verwaltungsvorschrift: administrative instruction/ regulation

Waldsterben: dying forests

Wirtschaftsstandort: economic location/ 'powerhouse' Germany

Wirtschaftswunder: economic miracle

zumutbar: reasonable or tolerable

1 Introduction

Why Study EU Environmental Policies in Subnational Regions?

The European Union (EU)[1] has established an impressive environmental policy. About 300 pieces of legislation are designed to regulate economic and other human activities and thereby tackle increasingly alarming environmental problems. EU environmental objectives include the setting of qualitative and quantitative standards (such as emission standards), the protection of species and areas of special interest (such as wild birds and habitats), and procedural frameworks which integrate environmental considerations into economic and other activities (such as environmental impact assessment). However, the effectiveness of most EU environmental policies has been hampered by the so-called implementation deficit whereby policy intentions on paper stand in stark contrast with policy reality.

The overall success of EU environmental policies depends on implementors' attitudes and resources on the ground. Yet, until recently their influence on the policies has been largely under-estimated and under-examined. Many EU practitioners and analysts have focused primarily on the creation of EU policies and the national governments' role in the policy-making process. Any assessment of the effectiveness of EU environmental policies, however, must include the evaluation of policy implementation within the Member States and their regions.

This research addresses the implementation deficit by contributing a perspective that moves away from the conventional state-centrist approach used among others by Sbragia (1996). It moves *sub*national regions[2] and their actors to the fore of EU environmental policy investigation and examines to what extent they shape EU environmental policies in practice.

The research compares two *sub*national regions, Scotland and Bavaria, and their implementation performances with EU environmental policies. It offers a *multi-layered implementation map* which guides the reader through the key government levels (EU, national, *sub*national levels), highlights obstacles and facilitators in the implementation path, and explains why implementation deficits occur. More importantly, the *map*

1

distinguishes between national and *sub*national government levels (or *layers*) and highlights the *sub*national regions' influence on the success (or failure) of EU environmental policies. By doing so, the *map* provides a more refined picture of the EU environmental policy 'reality'.

The following Chapter sets the study in its broader context. It begins by examining the relevance of environmental problems, the role of the EU in environmental politics, the problems of meeting EU environmental objectives, and the *sub*national regions' influence on EU environmental politics. The Chapter then formulates two key arguments and concludes with a structure outline of the book.

Considering the Wider Context of the Research Topic

Relevance of Environmental Problems

Concerns about pollution and environmental deterioration are relatively new. Apart from some isolated cases of water and air pollution caused by early industrial activity in the 19th Century, citizens in industrialised countries became aware of environmental deterioration only since the 1950s.[3] In Europe, especially, the negative impact of industrial activity and intensive farming not only shocked those who were directly affected by them; pollution and environmental deterioration also proved to know no frontiers and time limits. Acid rain caused by emissions of burning fuels such as coal frustrated citizens in all European states, so did the Chernobyl nuclear disaster of April 1986 which confirmed the perception that many environmental problems cannot be dealt with within national boundaries alone.

Environmental problems have been highlighted by alarming scientific evidence that responded and contributed to heightened public awareness. Environmental problems also received considerable attention from the media that highlighted signs of pollution and reached areas where 'green' issues had not been raised before. This new perception coincided with the wider societal adjustments in attitudes of the 1960s and 1970s that were initiated by the younger generation. The resulting New Social Movements did not entirely change society towards a 'post-materialist' or 'green' society, but they nevertheless succeeded in further raising public awareness on environmental issues.[4]

While shifts in perception were considerable and occurred in a relatively short period of time, they were not sufficient in providing the basis for effective solutions to problems of pollution and environmental deterioration. To date, the industrialised world has seen an abundance of environmental initiatives and legislation as well as recycling processes and the development of new 'clean' technologies to replace old heavy industries. These efforts, however, have borne limited success and the environment continues to deteriorate. To illustrate the point, the European Environment Agency report of 1995 states that Europe is far from moving towards a sustainable environment and is 'actually facing increasingly acute environmental difficulties'.[5] Despite optimistic findings by Jänicke et al (1996) who highlighted a decrease in certain pollution categories such as SO_2, CO and fertilisers, the 1980s and 1990s so far have shown slow progress in enforcing comprehensive and effective environmental measures. Policies and legislation undoubtedly have improved certain pollution problems. But the real driving force behind many success stories has been production changes from old heavy industries to new 'clean' technologies. These improvements have not eliminated threats and increases in pollution in other areas, neither have they solved the accumulative problem of pollution.

Three factors contribute to the rather cumbersome nature of environmental issues:

- environmental problems are complex and 'know no frontiers'
- their solutions are influenced by economic considerations
- environmental policies have to be processed within established and often inflexible societies and state systems.

With reference to the first factor, pollution and environmental deterioration constitute extremely complex problems. They affect other policy areas, 'know neither time nor territorial boundaries', are ever changing and involve a high degree of uncertainty. Moreover, environmental assets such as fresh air and rare species are difficult to value in economic terms.[6] These uncertainties hamper the tackling of environmental problems by conventional political and administrative means and require particular commitment and determination by politicians and citizens.

The second factor concerns the dominant perception that environmental considerations contradict economic success and material wealth. Embedded as a policy priority in post-war Europe, economic growth has been an essential objective for most citizens despite the post-materialist input of the 1960s and 1970s. Today, environmental issues are considered important, but in relation to other priorities such as job security, 'green' policies continue to be associated with inconvenience and costs.

Only in recent years, have attempts been made to reconcile environmental concerns and economic interests, though with limited success.

Finally, environmental matters are problematic because they often challenge existing societal and institutional structures. Past experience in the social development of the welfare state suggests that the process of accepting a new policy area and integrating it into the existing social, political and legal systems is difficult and slow. Institutional structures, which reflect and organise societal structures and preferences, are arguably slower in adapting to new demands. This lethargy is due in part to the administrators' perception that policy changes undermine their bureaucratic positions.[7] Environmental measures often require radical changes, for instance in the form of a new Ministry or a law which restricts certain economic activities. Therefore, environmental issues face similar problems to those faced by other new political issues (such as gender equality), namely the resistance and opposition of established interests. At the same time, environmental deterioration is intensifying and requires action which may stand in direct conflict with economic interests. The inherent complexity, the conflicting relationship with other interests and the urgency to act all make the environmental policy issue unique.

Relevance of the European Union

In Europe, environmental politics cannot be studied without focusing on the EU. The EU finds its roots in the European Coal and Steel Community (ECSC) of 1952 which signalled the beginning of a complex integration process involving quantitative (i.e. an increasing number of Member States and policies) and qualitative (i.e. intensifying policy commitments) developments. The EU has steadily acquired political and legal competencies that significantly affect every-day lives of EU citizens. Despite recent discussions on the principle of subsidiarity[8] and the European Commission's recent strategy to minimise the number of new policy proposals, the EU continues to shape fundamentally European politics related to the environment.[9]

Environmental policies constitute an area which has attracted considerable attention at the EU level. The objective to limit environmental deterioration and protect Europe's natural resources was formally adopted by the (then) European Community (EC) and its Member States with the Single European Act (SEA) in 1987. But even prior the SEA, from the 1970s onwards, the EC had begun to adopt environmental legislation as well as action programmes, initiatives and support funds. Today EU environmental policies are numerous and ambitious with an impressive number of Regulations and Directives regulating environmental matters.

Political and economic considerations have contributed to the adoption of environmental policies at the EU level. The first consideration concerns the apparent urgency of environmental problems described above. Particularly in times when public interest in environmental issues is high, politicians want to be seen as (and are expected to be) actively solving environmental problems at every opportunity. Inactivity could cost them the support of the electorate and would provide targets for criticism from political opponents. The EU presents itself as an ideal forum for concerted action, especially when problems of trans-boundary pollution are on the agenda. The EU provides a political and institutional framework that facilitates the realisation of environmental policy objectives and obliges Member States to follow their common commitments.

Economic interests, too, have forced Member States to establish the environment as a common policy area. In order to avoid economic imbalances caused by 'environmental dumping' whereby lax environmental standards in one Member State can attract economic activities away from another Member State, which has more stringent requirements, Member States have sought to harmonise their environmental policies and adopt a common environmental policy basis. This economic imperative has contributed towards the adoption of many EU environmental policies, among them the LCP Directive which sets common emission standards for large combustion plants, which otherwise would not have been adopted by EU actors.

The EU's 'historic commitment' to economic growth[10] has in many ways aggravated environmental problems in Europe. In particular, the effects of EU policies such as the Trans-European Networks and the Single Market have made EU environmental objectives even more pressing. In recent years, the EU has tried to integrate economic and environmental interests. EU documents such as the Task Force Report (1990) *1992: The Environmental Dimension* and the *Fifth Environmental Action Programme* bear the message of sectoral policy integration and sustainable development.[11] However, their influence on behavioural patterns and attitudes in the EU has to date been limited and not effective enough to bring about sustainable development in Europe.[12]

Meeting EU Environmental Policy Objectives

Considering the potential conflict between environmental and economic interests and the above-described inflexibility of societal and institutional structures, developments in EU environmental policy-*making* have been remarkably swift. However, when it comes to the actual implementation of EU environmental policies, progress has been markedly (and as expected, see above) slower and often disappointing. EU environmental policies have been confronted with severe obstacles to successful implementation. The EU environmental policy area in particular has shown an alarming and widening gap between 'good intentions on paper' and policy outcomes 'on the ground'.

The gap between EU environmental policy intention and reality is now recognised: several reports provide detailed evidence on the problems of implementation.[13] Until the early 1990s, however, considerable time was spent investigating EU environmental policy-making and the influence of Member State governments on policy formulation. This rather restricted perspective was partly due to the fact that the EC/EU had been preoccupied with the adoption of common environmental policies. In addition, implementation deficits were not as noticeable at the initial stages of the EU environmental policy process. However, with the adoption of an increasing number of environmental policies at the EU level, the contrast between policy objectives and policy outcomes has become more evident. In the light of increasing pressures to complete policy obligations and questions over the effectiveness and legitimacy of the EU level as a key policy-maker,[14] EU practitioners and researchers have started to re-adjust their focus by studying the problems of environmental policy implementation and by suggesting ways to consolidate environmental policy-making and implementation.[15] The European Commission, in particular, has highlighted the ever-increasing gap between EU environmental policy intention and the Member States' policy reality and has sought to close the gap through initiatives such as dialogue groups and partnerships.[16]

This research follows up the approach of EU environmental policy consolidation. It refines the picture of EU environmental policy implementation and highlights implementors 'on the ground'. In particular, the research identifies formal and informal conditions (or *determinants*, see below) on the ground that shape EU environmental policies during the implementation process.

Subnational Regions Matter

One obvious step towards consolidation is to focus on the latter stages of the policy process and study the problems of EU environmental policy implementation on the ground. Another step is to focus on the government level that plays a key function in putting most EU environmental policies into practice: the *sub*national level. EU environmental policies such as water and air quality Directives, nature protection Directives (e.g. Habitats Directive), or procedural Directives (e.g. Directive on environmental impact assessment) require implementation 'beyond' the national level – they affect the competencies and activities of *sub*national actors (i.e. actors at regional and local government levels) who have to implement and apply the policies and act in accordance with EU obligations. However, *sub*national actors do not exist in a vacuum; they are influenced and guided by formal-institutional structures as well as informal factors such as attitudes, priorities and relationships that exist within their *sub*national frameworks. These *determinants*[17] are not always compatible with EU environmental policies with the effect that their implementation is either inadequate or ignored completely. More importantly, many *sub*national *determinants* differ from the *determinants* that shape EU environmental policies in the Member States at large. Consequently, *sub*national regions feature implementation performances and problems with EU environmental policies that differ in many respects from their 'mother' states. The study of *sub*national regions is therefore essential when assessing the overall success (or failure) of EU environmental policies.

By and large, EU policy-makers and researchers have not paid enough attention to *sub*national regions and their role in the EU environmental policy process. This research brings the *sub*national regions and their actors to the fore of EU environmental policy investigation. It emphasises the importance of distinguishing between conditions at national and *sub*national levels and argues that the problem of the implementation gap in EU environmental policy cannot be fully grasped unless particular attention is paid to the 'unique' implementation conditions and performances inside the *sub*national regions.

Setting a Starting Point for Investigation

Bearing in mind the uniqueness and urgency of environmental issues; the significance of the EU as an important environmental policy-making level; the need to focus on the implementation stage of the EU policy process; and the importance of the *sub*national level in the EU environmental policy process, this book investigates the following two key arguments:

Argument (1): Formal *determinants* such as political-administrative structures as well as informal *determinants* such as policy priorities and relationships between actors influence EU environmental policy implementation on the ground. These formal and informal *determinants* are inter-related and cannot be studied on their own.

EU environmental policies have to be processed through various government levels until they reach the appropriate implementation level. In the process, policies are either altered, their implementation facilitated, or hindered by certain formal and informal conditions at the national and *sub*national levels. Conditions or factors that shape the process decisively can be described as *determinants*. To help identify implementation obstacles, the book distinguishes between seven formal and informal *determinant* categories that shape the EU environmental policy process in every *layer*. The formal *determinants* refer to:

- constitutional settings
- political-administrative structures and resources
- legal systems and instruments.

Informal *determinants* comprise:

- relationships between actors
- attitudes towards environmental protection and the EU
- policy-makers' priorities and strategies
- policy styles and practices.

Clearly, successful policy implementation depends upon favourable formal *determinants*. Yet, it would be a mistake to assess the policy process without taking into account attitudes and relationships between (and among) policy-makers and implementors. Even the most favourable formal political-administrative structure cannot guarantee satisfactory policy implementation if implementors are opposed to a policy or if relationships between policy-makers and implementors are not on constructive terms. On the other hand, relationships and attitudes do not develop free from any internal structures and legal frameworks; they are often the result of certain formal links and rules between political-administrative actors. Therefore, formal structures and processes significantly influence the attitudes and relationships between policy-makers and implementors.

The EU environmental policy process is shaped by formal and informal *determinants*, which are more complex than *determinants* in any other policy area. Not only do EU environmental policies face a longer process chain than national environmental policies (Directives in particular have to be transposed into national and *sub*national legislation and then require practical implementation); their objectives also have to fit in with the implementors' other (predominantly economic) priorities. In addition, EU environmental policies are influenced by the implementors' links with, and attitudes towards, EU policy-makers and, indeed, attitudes towards the European integration process in general. The research therefore takes account of the wider implications of environmental policies on other policy areas (e.g. economic competition and transport policies) and identifies EU-specific factors that influence EU environmental policy implementation.

Argument (2): *Sub*national regions and their actors play a central role in the EU environmental policy process. They shape the implementation of most EU environmental policies.

The second argument draws attention to the *sub*national regions and their role in the EU environmental policy process. In order to investigate *sub*national regions, however, it is necessary to define *subnationality*. A universal definition, which applies to all regions in the EU, is difficult to establish. EU Member States have evolved in different manners and feature diverse constitutional structures ranging from federal to unitary state systems. Consequently, *sub*national regions are diverse: they vary depending on their *sub*national regions' history, traditions and attitudes as well as their *embeddedness* within the national contexts.[18]

For the purpose of this research, *sub*national regions are defined as units immediately below the Member State level whose boundaries and identities are recognised by both the national governments and the EU. *Sub*national regions distinguish themselves from their 'mother' states by featuring 'unique mixes' of formal and informal conditions (or *determinants*). Many of these conditions develop independently within the regions and are, for instance, based on the regions' own history, culture and constitutional position. Other formal and informal *determinants* are influenced by the wider national and European contexts, i.e. events and circumstances that affect citizens not just within the regions but nation-wide or Europe-wide. Taken together, the resulting mixes of *determinants* are unique for each *sub*national region, they shape policy processes in their own way and should therefore be assessed in their own right.[19]

Two *sub*national regions have been selected for investigation which feature sharp differences as well as similarities: Scotland and Bavaria.

Until 1999 Scotland was an integral part of the United Kingdom's (UK) centralised state system and although Scotland was recognised as a 'nation' with a strong identity, it did not possess a government of its own.

Yet, it featured a number of formal and informal conditions that identified it as a region with a distinct political-administrative system. This distinctiveness, in turn, had an impact on the way EU environmental policies were implemented in Scotland. With the new Scottish Parliament and Executive, Scotland is currently in the process of developing its own independent public policy which is likely to have a further impact on EU environmental policy implementation in Scotland.[20]

In comparison, Bavaria has enjoyed considerable political-administrative autonomy for quite some time. As a federal state (Land) within the Federal Republic of Germany[21] Bavaria has shared political competencies with the Federal level and has been part of a multi-level, yet intertwined, state system. In such a system Bavaria has had ample opportunities to shape public policy and EU environmental policies in particular to suit Bavarian conditions and interests.

Referring to historical and constitutional discrepancies between the two regions, a former Secretary of State for Scotland once commented:

We are not Bavaria.[22]

Indeed, Scotland and Bavaria have differed (and continue to do so) in terms of their constitutional positions within the state systems, their policy priorities and geographical location.[23] At the same time, Scotland and Bavaria share common characteristics such as a strong 'national' identity and relatively large natural resources. In addition, both regions represent their interests in the EU more forcefully than most of their British or German counterparts. Overall, the comparison of both Scotland and Bavaria's similarities and differences promises valuable insights for the research; it helps define the *sub*national political-administrative systems and identifies those *determinants* that ultimately shape the implementation of EU environmental policies. Having stressed that Scotland and Bavaria are convenient objects for comparison, the research framework could easily be applied to other EU regions. Scotland and Bavaria simply serve as comparative examples that illustrate how *sub*national regions process EU environmental policies.

Research Structure

In order to address the above key arguments, this research applies a synthesis of analytical concepts that combines and refines the wider study areas of environmental politics, the politics of the European Union and the study of policy processes. The synthesis also introduces new elements to the subject: it highlights the *sub*national level in the investigation of EU environmental policy implementation and distinguishes between formal and informal *determinants* that shape the EU environmental policy process. A *multi-layered implementation map* illustrates how EU environmental policies are *filtered* through essentially three government levels and moves *sub*national regions to the fore by examining their influence on, and contributions towards, EU environmental policy outcomes.[24] The distinction between formal and informal *determinants* helps highlight facilitators and obstacles that lie in the implementation path of EU environmental policies.

The following Chapter, 'Towards a Multi-layered Implementation Map', outlines the synthesis in detail. It examines existing theories related to the research topic and draws up a *map* that best encompasses the EU environmental policy process and guides the reader. Starting with the first *layer* of the implementation *map*, Chapter 3 focuses on the EU level: it introduces the complex EU environmental policy-making process, outlines the key EU environmental policy objectives and, finally, contrasts EU policy intentions with the most common problems and insufficiencies during policy realisation. Chapter 4 describes the environmental policy and politics of the UK and Germany and assesses to what extent formal and informal *determinants* at the national level influence the (non-) implementation of EU environmental policies. It concludes that the study of national *determinants* is not enough to explain fully the implementation deficit. Chapter 5 contributes the *sub*national level to the study of EU environmental policy implementation. It establishes the Scottish and Bavarian political-administrative systems, highlights their environmental politics and policies, and identifies 'uniquely' Scottish and Bavarian *determinants* that influence EU environmental policy outcomes. The Case Study featured in Chapter 6 investigates the process and problems of EU environmental policy implementation in detail. It focuses on one particular piece of legislation, the Directive on environmental impact assessment (85/337/EC), and compares its implementation in the national and *sub*national *layers*. Chapter 7 concludes with a resume of the research; it addresses the key arguments again in the light of the evidence and assesses the usefulness of the *multi-layered implementation map* for further investigations.

Overall, this research seeks to explain why - despite good intentions - EU environmental policies fail to be properly implemented and why the EU remains a long way from its objective of sustainable development. It does not suggest a new theory that predicts policy outcomes; neither does it propose ultimate solutions to the implementation deficit. Rather, it offers more refined insights into EU environmental policy implementation practices and contributes towards a better understanding of the implementation deficit. The new evidence could in turn contribute towards closer co-operation between the government levels that could, ultimately, improve implementation practices of EU environmental policies.

Notes

[1] With the Treaty on European Union (TEU) of 1992 the Member States included two policy 'pillars' (Justice and Home Affairs, Common Foreign and Security Policy) which were incorporated to a certain extent into the main treaty framework with the Amsterdam Treaty of 1997. The whole 'temple' construction, however, was named European Union (EU) which is the official term since the ratification of the TEU on 1. November 1993. Environmental legislation is adopted under the European Community (EC) pillar. The term EC (which was accepted as the general term before TEU ratification) replaced the European Economic Community (EEC). All these changes make the correct use of terms difficult. In this book the term EC is applied in the context of activities conducted before 1. November 1993, the term EU is applied in general but more specifically in relation to activities after 1993. In cases where policies and activities cannot be put into one of the two categories the author uses both terms in conjunction, i.e. EC\EU.

[2] The author is well aware of the negative connotation the terms *subnational* and *region* may have for many Scots and Bavarians. In their defence, the terms were chosen in the context of the EU policy implementation process and do not reflect national identities and other (ideological) factors.

[3] The London 'Great Smog' in December 1952, which caused over 4,000 deaths, was one of the first major pollution incidents to receive wide attention. See Lean (1995).

[4] For further information on New Social Movements see Inglehart (1990), Roth and Rucht (1991), Raschke (1988). For a critique on New Social Movements see Koopmans (1995).

[5] See Mann (1996). Previously, the *EU Fifth Environmental Action Programme* pointed out '[s]ome disquieting trends' which, 'if not satisfactorily contained, could have significant negative consequences for the quality of the environment' in Europe. Trends included a 20% increase in EC carbon emissions by 2010 and a 13% increase in municipal waste over a 5-year period. See *O.J.* (19.93).

[6] For a discussion on attributing an economic value to environmental assets see, for instance, Goldin and Winters (1995). See also *The Economist* (1998).

[7] Among other implementation researchers, Pressman and Wildavsky (1974) describe the administrators' reluctance to adjust, or give up, their positions for the sake of new policies.

[8] The EU defines subsidiarity as locating decision-making at the 'most appropriate level'. Ideally, decisions should be made as closely as possible to the citizens, i.e. at the 'lowest' government level. However, where the 'lowest level' cannot adequately address policy issues, a 'more appropriate' level (e.g. the EU) should act. For a summary and discussion of the EU principle of subsidiarity, see Scott et al (1994).

[9] Bomberg (1999) highlights the Commission's 'do less but do it better' strategy in EU

environmental policy.

[10] The term 'historic opportunity' is used by Baker (1997).

[11] The Brundtland Report (1987) defines 'sustainable development' as a development that meets the needs of the present without compromising the ability of future generations to meet their needs. Similarly, the EU defines 'sustainable development' as a 'policy and strategy for continued economic and social development without detriment to the environment and the natural resources on the quality of which continued human activity and further development depend' (p.12). See *EU Fifth Environmental Action Programme*. 'Sustainability' can be defined as the long-term (final) objective in achieving a complete merger of economic, social and environmental developments.

[12] For a discussion of sustainable development in the EU and case studies on Member States and regions' experiences with sustainable development, see Baker et al (1997).

[13] Among others see the Institute for European Environmental Policy (1993), the Commission's *11th Annual Report* (1994) and Krämer (1992 and 1997).

[14] Merkel (1999) discusses in more detail the questions of EU legitimacy and effectiveness.

[15] A similar approach of consolidation applies to policy objectives of the Single Market and, indeed, the future Economic and Monetary Union (EMU).

[16] In the *EU Fifth Environmental Action Programme* the Commission introduced three dialogue groups: the General Consultative Forum (since 1997 European Consultative Forum on the Environment and Sustainable Development, in short Forum), the Implementation Network and the Environmental Policy Review Group. The 'partnership' between policy-makers and implementors at EU, national and *sub*national levels was confirmed in the Commission's *Interim Review* (1994).

[17] The author uses the term *determinants* to describe factors that decisively influence the implementation process of EU environmental policies. *Determinants* can facilitate these policies, alter them, or hinder their proper implementation. For a more detailed definition and justification of the term see below and Chapter 2.

[18] Even within the UK there are various forms of *regions*. In particular, Scotland possesses county councils similar to councils in England and Wales, yet Scotland as a whole is often treated as one region. See Rhodes (1997).

[19] For another definition of *sub*national regions see Udo Bullmann's typology of regional organisation in Jeffery (1997). He distinguishes between regions belonging to either classic unitary states, devolving unitary states, regionalised unitary states, or federal states. For a list of definitions (including the Commission's own definition) see Edye (1997). Finally, Bomberg and Peterson (1996) categorise *sub*national regions along a scale of constitutionally strong and weak SNAs (sub-national authorities).

[20] The field research for this book was conducted between November 1994 and February 1999.

[21] The constitution of the Federal Republic of Germany (FRG) of 1949 guaranteed Länder autonomous powers for Bavaria. Until October 1990 post-war Germany was divided into the Federal Republic and the German Democratic Republic (GDR). While the GDR joined the Warsaw Pact and Comecon, the Federal Republic proceeded with European integration and joined NATO. Following the 'velvet revolution' and collapse of the Eastern Bloc, the GDR ceased to exist in 1990 and five 'new' Länder joined the Federal Republic. While the Federal Republic still exists as such, it is now common practice to refer to it as Germany. The author follows this general trend by omitting 'Federal Republic'. Where it is deemed necessary, however, the author emphasises the term 'Federal'.

[22] Former Secretary of State for Scotland Ian Lang, at the launch of the 'Scotland Europa' office in Brussels. Quoted in *The Herald* (1992).

[23] Scotland represents a *sub*national region at the periphery of the EU with its own distinct characteristics and problems (e.g. disadvantage in terms of trading links with the EU 'core', comparatively large natural resources but a keen interest in economic development), while Bavaria is geographically at the 'heart of Europe' and faces problems of a different kind (e.g. economic and environmental pressures caused by immigration and North-South/ East-West travel patterns). For a detailed investigation of peripheral regions and EU environmental policies see Baker et al (1994).

[24] By focusing on *sub*national regions, the book borrows insights from the 'multi-level governance' approach which emphasises the importance of a 'third tier' (i.e. *sub*national government tier) in the EU policy process. For 'multi-level governance' analyses see Hooghe (1996); Marks et al (1996); Jeffery (1996a and 1996b).

References

Baker, S. et al (1994), *Protecting the Periphery. Environmental Policy in the Peripheral Regions of the EU*, Frank Cass, London.

Baker, S., Kousis, M., Richardson, D. and Young, S. (eds) (1997), *The Politics of Sustainable Development. Theory, Policy and Practice within the European Union*, Routledge, London.

Bomberg, E. and Peterson, J. (1996), *Decision-making in the European Union. Implications for central-local government relations*, Joseph Rowntree Foundation, York.

Bomberg, E. and Peterson, J. (1999), *Decision-making in the European Union*, MacMillan, Basingstoke.

Commission (1993), *5ᵗʰ Environmental Action Programme Towards Sustainability*, Official Journal No C138.

Commission (1994), *Eleventh Annual Report to the European Parliament on monitoring the Application of Community Law - 1993*, (Section G. Environment), Official Journal No C154.

Commission (1994), *Interim Review*.

Edye (1997), *Regions and Regionalism in the European Union*, European Dossier 38, University of North London.

Goldin, I. and Winters, L.A. (eds) (1995), *The Economics of Sustainable Development*, Cambridge University Press, Cambridge.

Hooghe, L. (ed) (1996), *Cohesion Policy and European Integration. Building Multi-level Governance*, Clarendon Press, Oxford.

Inglehart, R. (1990), *Culture Shift in Advanced Industrial Society*, Princeton University Press.

Institute for European Environmental Policy (1993), *The State of Reporting by the EC in Fulfilment of Obligations contained in EC Environmental Legislation*, London.

Jänicke, M. et al (1996), *Umweltpolitik der Industrieländer. Entwicklung-Bilanz-Erfolgsbedingungen*, edition sigma, Berlin.

Jeffery, C. (1996a), 'Towards a 'Third Level' in Europe? The German Länder in the European Union', *Political Studies*, vol.44, No.2, pp.253-266.

Jeffery, C. (1996b), 'Sub-National Authorities and European Domestic Policy', *Regional and Federal Studies*, vol.7, No.3, pp.204-219.

Jeffery, C. (ed) (1997), *The Regional Dimension of the European Union. Towards a Third Level in Europe?*, Frank Cass, London.

Koopmans, R. (1995), *Democracy from below. New Social Movements and the political system in West Germany*, Westview Press, Oxford.

Krämer, L. (1992 and 1997), *Focus on European Environmental Law*, Sweet and Maxwell, London.

Lean, G. (1995), 'Where did all the fresh air go?', *The Independent on Sunday*, (The Sunday Review, 5. March), pp.4-9.

Mann, M. (1996), 'Widespread condemnation of action programme review', *European Voice*, (Survey: Environment, 30 May - 5 June), p.17.

Marks, G. et al (eds) (1996), *Governance in the EU*, Sage, London.

Merkel, W. (1999), 'Legitimacy and Democracy. Endogenous Limits to European Integration', in Anderson, J. (ed) *Regional Integration and Democracy*, Rownan and Littlefield.

Pressman J.L. and Wildavsky A. (1974), *Implementation: How great expectations in Washington are dashed in Oakland*, Los Angeles, University of California.

Raschke, J. (1988), *Soziale Bewegungen. Ein historisch-systematischer Grundriss*, Campus.

Rhodes, R.A.W. (1997), *Understanding Governance. Policy Networks, Governance, Reflexivity and Accountability*, Open University Press, Buckingham.

Roth, R. and Rucht, D. (eds) (1991), *Neue Soziale Bewegungen in der Bundesrepublik Deutschland*, Bundeszentrale für politische Bildung, Bonn.

Sbragia, A. (1996), 'Environmental Policy: The Push-Pull Policy-making' in Wallace, H. and Wallace, W. (eds) *Policy-making in the European Union*, (pp.235-255.

Scott, A. et al (1994), 'Subsidiarity: A 'Europe of the Regions' versus the British Constitution?', *Journal of Common Market Studies*, vol.32, No.1, pp.47-67.

Task Force Environment and the Internal Market (1990), *'1992' The Environmental Dimension*, Report.

The Economist (1998), 'An invaluable environment', 18. April, p.105.

The Herald (1992), 'Opening night for Scotland's voice in Europe', 27 May, p.1.

World Commission on Environment and Development (1987), *Our Common Future*, (Brundtland Report), Oxford University Press, Oxford.

2 Towards a Multi-Layered Implementation Map

Introduction

Processing findings within a conceptual framework facilitates research and contributes to a greater understanding of the subject area. This Chapter introduces theories and concepts relevant for the study of EU environmental policy implementation. It assesses their usefulness and extracts relevant components which will form part of the analytical framework. The Chapter concludes with a *multi-layered implementation map* which includes key insights from existing concepts and adds new elements to the study of EU environmental policy implementation.

Studying Policy Implementation

> If Implementation Works - Clap!

The above comment by Richardson (1996) indicates that the implementation of policies should not be taken for granted. In fact, over the years many analysts have attempted to address the question why policy commitments often fail in practice. Implementation analysts have complained that most studies concentrate on the study of policy-making while the subsequent implementation of policies has been neglected as a focus for research.[1] Calls for more implementation studies commenced as early as the 1970s and the criticism over the neglect of implementation as a scientific focus still continued in the 1990s. Nevertheless, by the mid 1980s an impressive list of about 90 implementation studies was produced by O'Toole (1986), which suggests that this study area is not as exotic as many people believe. What is missing is a comprehensive analytical framework that can be applied to a wide range of possible scenarios of policy implementation. Some analysts have attempted to establish all-encompassing theories, while others have claimed that they have found the most important and decisive factor that determines policy implementation.

While all analysts have contributed valuable insights to the discipline, none of them provides a framework which adequately addresses the complexity of the EU environmental policy scenario. The following review outlines relevant implementation concepts and then extracts analytical tools that contribute towards a more suitable analytical framework for this research.

Pressman and Wildavsky's Implementation Theory

One of the first major studies on policy implementation remains in many ways the most valuable contribution to this study area: Pressman and Wildavsky's *Implementation* of 1974. Even though their study describes the implementation efforts of the Economic Development Administration Program in Oakland, California (USA),[2] many of Pressman and Wildavsky's conclusions can be applied to the EU context.

At the outset both authors assumed that the implementation of a well-prepared and popular program such as the Oakland Program would be accomplished with no delay and difficulties. However, the subsequent failure of the 'Program' showed that a policy - no matter how perfect it appears to be in its formulation - does not necessarily lead to its complete, satisfactory implementation. Surprised by the failure of an apparently popular policy, Pressman and Wildavsky decided to identify the reasons for non-implementation.

Following their investigations, Pressman and Wildavsky pointed out that there are *various forms of policies* applied by political actors. Sometimes policies represent a statement of intention, at other times they describe a certain standpoint or behaviour. However, if a policy contains a certain objective to be achieved, the implementation of that policy cannot be ignored. Applied to the EU, environmental policies clearly set out policy objectives that require implementation at a later stage. Once adopted, EU environmental policies are legally binding and Member States are expected to follow their policy obligations. EU environmental policies belong to the latter category of Pressman and Wildavsky's policy forms. In order to gain a full picture of an EU environmental policy, it is therefore essential to investigate 'beyond' policy statements and consider their implementation on the ground.

A policy followed through to its actual implementation has a *starting point* and an (ideal) *end point*. What occurs between starting point and end point can be calculated or predicted as accurately as possible; nevertheless a policy at the starting point is a mere prediction or hypothesis at that time. The implementation of new policies always requires new legislation, funding and coordination, as well as co-operation at various levels and administrative changes. These factors indicate a variety of obstacles that have to be overcome in order to reach the policy goal. In addition, dealing with one obstacle may influence another obstacle. A 'program', for example, is a system of policy stages that are related and dependent on each other. If one stage in the 'program' chain faces difficulties, the following link is very likely to be affected. Pressman and Wildavsky labelled potential obstacles in the process *decision points* and *clearances*. The number of decision points and clearances determines the likelihood of successful policy implementation; the more actors involved in clearances, the more difficult it is to implement a policy successfully.

The sheer number of implementors, however, does not always determine policy outcomes.[3] Nevertheless, Pressman and Wildavsky's decision points and clearances are useful in so far as they illustrate potential obstacles in the EU environmental policy implementation process. With the EU as a *supra*national government level, the EU policy process is obviously more complex than national policy processes. The EU policy process involves not only an additional government level, it also involves a wider range of actors and interests than in the case of national processes. The likelihood of an EU environmental policy being blocked by an obstacle is therefore quite substantial. In addition, environmental policies tend to be complex and often affect economic, social and other policy areas. Consequently, representatives of other interests tend to get involved in the process and place obstacles in the implementation path of EU environmental policies.

Following Pressman and Wildavsky's argumentative line further, the *time span* plays an important role in the implementation result. The more time is envisaged for a 'program', from policy initiation to desired implementation, the more difficult it becomes to reach the end point. Too many changes can occur in a period of time - circumstances, opinions, scientific evidence etc. can change, so can goals or policy objectives themselves. In the case of EU environmental policies, most policies are formulated as Directives which contain implementation deadlines usually effective three years after adoption. While three years may be necessary to adjust national systems and constitute a short period to put certain environmental obligations into practice (such as setting up offices, adjusting national legislation, requiring polluters to invest in clean technologies), in political terms three years can be a long time in which actors tend to forget policy obligations and commitments. The large time span between policy adoption and implementation deadline can therefore have a negative impact on the effectiveness of an EU environmental policy.

Successful implementation also depends upon *feasibility*, according to Pressman and Wildavsky. If aspirations are set too high, the objective is unlikely to be accomplished. Goals should, therefore, be within reach. If a policy goal had not been reached in a previous 'program', the enthusiasm for a subsequent 'program' may be negatively affected. The implementation and success of previous policies can therefore affect policy-making significantly. As far as EU environmental policies are concerned, views differ on the feasibility aspect. According to environmental NGOs and some Member States (in particular Denmark, Germany and the Netherlands),[4] most policy goals are set at the lowest common denominator level,[5] while many industrial lobbyists and other Member States consider the policy goals as too ambitious and unrealistic. Pressman and Wildavsky miss the point that the feasibility of EU environmental policies depends upon the perception of implementors: if policy goals are seen as unattainable, implementors are unlikely to put much effort into the policy. Pressman and Wildavsky neglect the importance of perceptions but their observation is correct concerning the impact of previous policy performances on further policy implementation: the success or failure of EU environmental policies influence attitudes at a later stage when other policy goals are supposed to be implemented.

Pressman and Wildavsky attended to *actors and their attitudes* in a separate section of their analysis. If the motivation and support towards a policy is high, implementation is completed without problems or delays. If the perception is critical or even negative, implementation is hindered considerably. Lack of interest in a policy can delay implementation moderately and should not be ignored in the investigation. Attention should also be paid towards the question of priorities and competing interests that have an impact on the policy implementation outcome. If the policy is perceived as a major priority, its implementation is most likely to be accomplished. If, however, the policy has to compete with other interests on the practical implementation level, the actual implementation is most likely to fail. Whether or not an actor is involved in policy-making is also important for implementation. If an actor is instructed to implement a policy but had no say in the shaping of the policy, this person tends to fulfil his\ her implementation tasks with less enthusiasm or even with reluctance. Pressman and Wildavsky stressed that some implementors may well agree with the substantive aim of a policy but are unable or unwilling to implement the policy in detail. Possible reasons include simultaneous commitments to other projects, dependence on other actors who are less committed towards the policy, and incompatibility with existing policies or procedures.

A similar behavioural pattern applies to implementors of EU environmental policies. In fact, the implementation of EU environmental policies is heavily dependent upon attitudes towards the EU and environmental objectives. Implementors are required not only to accept policy decisions coming from the EU level, they are also required to accommodate environmental objectives which may conflict with their existing (economic) priorities. The gap between EU policy-makers and implementors represents another handicap for the implementation of EU environmental policies. Despite the Commission's efforts to involve implementors in the policy-making process, the majority of implementors at the grass-root level is excluded from the negotiating process and considers EU environmental policies as 'instructions from above'. According to Pressman and Wildavsky, this non-involvement can only have a negative impact on EU environmental policy outcomes.

Overall, Pressman and Wildavsky describe the complex relationship between policy-making and implementation and highlight the gap between the two policy process stages. Their study demonstrates that favourable conditions during policy-making (support for a policy, allocation of resources, etc.) are not necessarily met with successful policy implementation. Certain factors such as attitudes and the time span can block an implementation chain that may then result in a policy's failure. While Pressman and Wildavsky's findings constitute valuable material for this research, their study also shows flaws (such as the feasibility argument) and neglects areas which still require consideration. In particular, critics such as Sabatier and Elmore (mentioned below) have pointed out that Pressman and Wildavsky's concept of policy processes is too linear and hierarchical. It concentrates on single 'top-down' instructions 'from above' while neglecting the influence of 'grass root' actors in shaping policies throughout the policy process. The following studies attempt to construct more comprehensive implementation theories that address the flaws of Pressman and Wildavsky's concept.

A Discourse of Policy Implementation Studies

Discussions over the 'right' implementation approach have occupied analysts such as Jordan (1996 and 1996a); Hill and Weissert (1995). *Top-down* analysts such as Pressman and Wildavsky have started their investigation with the presentation of a policy decision and then addressed questions that concern the consistency between the policy objective and policy reality. However, *top-downers* have often neglected pre-decisional factors that determine the policy outcome considerably. In many instances, policies under *top-down* investigation have appeared to have 'come out of the blue' or appeared to have been created by policy-makers without prior consultation and co-operation with interested parties.[6]

Bottom-up analysts such as Elmore (1979) have focused instead on the complex, reciprocal relationship between both levels and stressed the continuous learning process between policy-makers and implementors. Elmore refused to accept the hierarchical structure of *top-down* analysis. This approach, in his words *forward mapping*, implies a linear downward instruction that is neither influenced nor challenged by implementors. However, implementors at the lower policy levels influence policy outcomes considerably. It would therefore be inadequate to analyse policy implementation by concentrating on the transposition of policy instructions from above without considering attitudes and conditions at the bottom. Elmore argued further that a *backward mapping* approach would not only benefit analysts in their analysis. Applied in practice, backward mapping would also improve policy outcomes. Policy-makers should assess implementors' abilities, attitudes and resources first before they formulate a policy. Considering the 'real' conditions at the bottom ensures successful accomplishment of a policy.

In the light of an increasing gap between EU environmental policy objectives and their implementation, Elmore's idea of *backward mapping* is attractive for EU practitioners and analysts as it suggests a solution to the problem of implementation deficit.[7] However, one major problem arises with Elmore's concept. His approach does not highlight the discrepancies between policy *intention* and policy *reality*. Rather, it adjusts the research perspective to the implementors' level without considering the policy-makers' legitimate position in the process and their hopes and expectations. Policy-makers derive their legitimacy from direct elections (or other selection procedures), they respond to public demands and (urgent) problems and adopt policies on behalf of their constituents. It would therefore be a mistake to neglect their importance in the overall policy process. Elmore's concept is useful as it brings implementors' interests and behaviours to the fore, but in the case of the EU environmental policy implementation deficit, Elmore's *backward mapping* is of limited use because it neglects the EU policy-makers' role in the overall process.

In an attempt to provide a more comprehensive framework for the study of policy implementation, Sabatier and Mazmanian (1980 and 1981) produced a concept which is almost at breaking point because of the weight it carries. Sabatier and Mazmanian established three broad categories of factors which shape the implementation process:

- tractability of the problem(s) being addressed by a statute
- the extent to which the statute coherently structures the implementation process
- non-statutory variables affecting implementation.

The lists of items under these categories are not always plausible as some items occur in all three categories. Particularly, a clear line cannot be drawn between the first category (*tractability*) and the third category (*non-statutory variables*). Moreover, Sabatier and Mazmanian's flow chart model of the policy process assumes that a policy problem suddenly occurs without prior influence of political variables. Despite these shortcomings, Sabatier and Mazmanian's study is useful in so far as they suggest a checklist of ideal conditions for policy implementation which helps predict policy outcomes. In an 'ideal scenario', legislation should outline clear and consistent objectives, demonstrate causal links between problem and problem-solving objectives, and specify responsibilities in the implementation process. Further, implementors and target groups should possess the necessary means to accomplish policy objectives and should be committed towards the policy consistently over the whole implementation period.

Applied to the research, outlining an EU environmental policy that is clear and consistent for all involved is a difficult (if not impossible) task considering the EU's vast diversity in terms of languages, legal traditions and political priorities. Further, demonstrating a causal link between environmental problems and EU environmental policy objectives is enormously difficult since most environmental objectives are complex, involve long-term periods and are inter-connected with other policy areas. In many cases, responsibilities are not specified in the legislation and EU Member State governments tend not to supervise adequately the implementation of EU environmental policies. This lack of clarity can result in confusion and is occasionally used to avoid unwelcome policy obligations. In addition, financial and administrative resources required for the implementation of EU environmental policies are generally not specified in the legal texts to allow for national and *sub*national variances and diversity. However, resources allocated at a later stage often prove to be insufficient for effective implementation. In addition, the commitment of implementors is often missing due to a lack of consultation and involvement at the policy-making stage. Considering Sabatier and Mazmanian's criteria, EU environmental policies face serious implementation difficulties indeed. The EU environmental policy implementation 'scenario' is therefore much further from 'ideal' than are national policy implementation 'scenarios'.

Sabatier (1986 and 1998) offered another all-comprising theoretical framework which merges various approaches and encompasses all directions and influences in the process of policy change. In order to limit the enormous complexity of his framework, Sabatier proposed the categorisation of *advocacy coalitions*, i.e. groupings who share sets of beliefs and seek to realise common goals in a policy system. His *advocacy-coalition* model is useful in so far as it takes into account (conflicting) interests and influences of various factions within policy systems as well as the role of political-administrative *policy brokers*. On the other hand, his model implies that influential factors - *relatively stable system parameters* (such as socio-cultural values and basic constitutional structures) and *events external to subsystem* (such as changes in socio-economic conditions and government changes) - are on the sideline of the policy process. According to the model, *advocacy coalitions* seem to be affected by these influential factors but remain outside their framework. However, actors such as *policy brokers* derive their positions from the *system parameters* and are an integral part of them. Sabatier's model is misleading at this point. Nevertheless, his *advocacy coalitions* are useful for the EU environmental policy context. They highlight the tensions between environmental coalitions and their opponents who fear the costs of EU environmental policies. Moreover, Sabatier's model describes the Commission's problematic position as a *policy broker* in the EU environmental policy process which negotiates between conflicting interests. His model helps explain the Commission's behaviour towards poor implementation performances; playing the role of a policy *broker*, the Commission is often overly tolerant and compromising towards 'bad' implementors.

The above studies describe policy processes in general. Other analysts have preferred to focus on individual aspects which they consider as the key to implementation problems. Some analysts have drawn attention to the evaluation of policy implementation and its effects on policy-making,[8] while others have been interested in the clashes of two or more policies and their hindering impact on the policies' implementation.[9]

Thompson (1984) concentrated on the interdependent relationships between policy-makers and implementors which determine policy outcomes.[10] He distinguished between *pooled, sequential,* and *reciprocal* interdependencies and outlined the advantages and disadvantages of each relationship category.[11] EU relationships can be described as *sequential* and *reciprocal* interdependencies involving a multitude of actors, channels and directions. The *sequential* and *reciprocal* categories are useful in so far as they highlight the complexity and interconnectedness of the EU process. They help identify the advantages of complex EU relationships (such as feed-back on policies and the participation of grass-root actors), as well as downfalls (such as delays and possible disagreements) in the EU policy implementation process. While Thompson's focus on policy process relationships is useful in highlighting weak links between EU actors, his concept neglects other essential implementation aspects such as the influence of policy instruments (or tools) on policy outcomes.

Ingram and Schneider (1990) stressed the importance of studying policy tools and proposed the framing of *smarter statutes* that forestall implementation problems. They distinguished between four statutory categories:

- strong statute
- grass roots statute
- support building statute
- Wilsonian statute.[12]

The *Wilsonian statute* is of particular interest because it resembles EU Directives: it combines policy goal specificity with discretion over the ways and means in achieving the policy goal.[13] Once a *Wilsonian statute* has been issued, it is generally perceived as a 'depoliticised responsibility' for professional administrators. The statute leaves an essential part (i.e. ways and means) of policy implementation to the implementors' discretion, at the same time policy-makers lack effective control over the realisation of policy commitments. The resulting gap between policy-makers and implementors, described by Ingram and Schneider, is evident in the EU policy process: once EU policy-makers issue a Directive, they often lack effective control powers over implementors at the national and *sub*national levels. The evident gap between EU policy-makers and implementors caused by the statute form makes the proper transposition of a policy difficult.

Berman (1980), too, focused on statute forms and their influence on policy implementation. He established two broad statute categories, which can be located at two ends of a scale: statutes are either accomplished according to *programmed* implementation or *adaptive* implementation. *Programmed* statutes constitute well-defined policies which allow only limited discretion for implementors, provide sufficient monitoring and control powers for political decision-makers, outline incentives and disincentives, anticipate problems, and are formulated as clear and precise as possible. *Adaptive* statutes, on the other hand, provide only general policy objectives with maximum discretion for implementors while policy-makers resume a passive role. In contrast to *programmed* statutes, *adaptive* statutes are open to modifications and revisions during the implementation process. More importantly, *adaptive* statutes lack a clear line between policy-making and implementation as far as policy decisions and actors are concerned.

Like other statute analysts, Berman suggested that the most appropriate policy tool should be chosen to suit the policy situation. Since policies tend to be complex constructions consisting of several political messages and containing legal, administrative and resource provisions, a combination of statute ingredients should be applied where appropriate. While the 'pick and mix' approach is sensible as it allows for complex circumstances and objectives, EU environmental policy outcomes have shown that complex statutes, carefully formulated to contain all aspects (and interests) of the policy matter, do not necessarily result in successful implementation. Following Berman's concept, most EU environmental Directives are neither *programmed* nor *adaptive* statutes but contain a complex mix of statute ingredients. In practice, this mix does not prevent the EU from failures in the environmental policy area. Mixed EU statutes are sensible but cannot eliminate implementation problems such as misunderstandings over requirements, lack of supervisory power and lack of discipline. Therefore, the study of statutes forms helps identify EU Directive characteristics (i.e. discretionary and regulatory elements) that contribute significantly towards policy outcomes. Studying statutes on their own is useful as long as researchers keep in mind that they do not provide a complete picture of EU environmental policy implementation problems.

The above findings reflect the state of policy implementation studies and their usefulness for this research. In order to complete the analytical framework, the following Sections introduce the dimensions of environmental politics and EU politics.

Adding the Environmental Dimension

Chapter 1 introduced the environmental policy area as an area which has no counterpart in terms of complexity, urgency and inter-relationships with other policy areas. For the study of EU environmental policy implementation one of the key characteristics of the environmental policy area requires particular attention: the inter-relationship between economic and environmental considerations.

Offe's *subsystems* model (1984) is a useful starting point in establishing the relationship between environmental and economic interests. Offe investigated the functioning of a state system from a post-Marxist perspective and highlighted tensions and conflicts between economic (or 'capitalist') interests and other interests which may exist already or arise in a developed 'capitalist' society. According to Offe, modern society evolves around three *subsystems*: the *economic subsystem*, the *normative* (legitimisation) *subsystem* and the *political-administrative subsystem*. Put simply, the *economic subsystem* stands for economic or 'capitalist' interests which in many cases exclude 'moralist' (or environmental) considerations. The second *subsystem*, the *normative subsystem*, represents 'public morale' or public pressure. This pressure may be targeted against 'capitalist' interests causing tension between the two *subsystems*. The third, *political-administrative*, *subsystem* provides the framework within which both diverging interests can negotiate and secure as many interests as possible. Similar to Sabatier's *policy broker*, Offe's *political-administrative subsystem* plays the role of a mediator or referee, seeking to accommodate both sides.[14] According to Offe, the domination of one *subsystem* would inevitably lead to instability if not self-destruction of the 'capitalist' state as a whole. Therefore, the balancing act between the *subsystems* is vital. Of course, by mediating between economic and 'moralist' interests, the *political-administrative subsystem* has an interest of its own, namely to survive and confirm its own position. It is dependent upon the *normative subsystem's* approval as well as the support of the *economic subsystem* which is essential for financing the *political-administrative subsystem* and, in fact, the functioning of the state system in general. Crises still occur despite the balancing, self-regulatory nature of the developed 'capitalist' state: the *political-administrative subsystem* is often unable to cope with the tensions. The reason for this dilemma lies in the dual interest of the *political-administrative subsystem* to please both the *economic* and the *normative subsystems*.

Offe's model is useful because it highlights wider 'macro' conditions that shape the formulation and implementation of environmental policies. His model is also useful in so far as it highlights the tensions between two conflicting interest groups as well as the political-administrative actors' efforts to mediate between the factions. In the case of the EU, environmental interest groups often clash with representatives of economic interests over EU environmental policies. The Commission, in turn, is constantly seeking to reach compromise solutions that accommodate environmental and economic interests.[15] In doing so, the Commission itself has to reach a compromise solution as it is subdivided into sectoral Directorates-Generals (DGs), which constantly guard 'their' policy interests.[16]

There are weaknesses in Offe's model which limit its application to the study of modern environmental politics. Firstly, Offe's model does not explicitly address the idea of sustainable development, i.e. the concept to reconcile economic, social and environmental developments. True, environmental and economic interests often stand in conflict with each other. On the other hand, sustainable development is reflected in a number of private and public sector initiatives which seek to integrate environmental and economic concerns.[17] Secondly, Offe neglects the economic cycle and its impact on environmental politics. In times of economic recession, relations between environmental and economic interest groups are likely to be tense and environmental objectives are pushed to second place in priorities. In times of economic prosperity, environmental concerns can be pursued with fewer economic obstacles. In other words, during periods of economic growth and prosperity, citizens are more receptive to environmental concerns and 'can afford' environmental policies, while in times of recession; citizens tend to concentrate on economic (and related social) issues and 'cannot afford' environmental policies. Many observers describe this correlation as a 'paradoxical' relationship between economic and environmental interests.[18]

Finally, modern politics is not easily divided into Offe's three *subsystems* and constrained into a single-state framework. In the case of the EU, a line between environmental and economic interest groups cannot be drawn at all times, considering the complexity and inter-connectedness of EU actors and their interests. The EU environmental policy process involves fifteen Member States, several government levels and inter-related policy areas. While Offe's single-state scenario is difficult to apply to the EU context, his macro perspective is nevertheless useful as it emphasises the (potential) conflict between economic and environmental interests of the EU and the dilemma to try and bring the two interests together. The dilemma is particularly evident when it comes to the formal and practical implementation of EU environmental policies. The following Section specifically deals with the EU dimension.

Adding the EU Dimension

The EU provides fertile ground for research and has attracted many analysts who have tried to understand and explain the complicated EU policy process. However, in contrast to studies on the EU bargaining and decision-making process, which exist in abundance, the number of studies on EU policy implementation is somewhat limited. One obvious explanation lies in the initial preoccupation of the EU to produce policies, while their implementation at the national and *sub*national levels has been outside EU analysts' field of vision. Only in recent years, in the light of an increasing number of unresolved EU policies, have EU policy-makers and analysts recognised the widening gap between EU policy objectives and policy 'reality' which is now causing considerable concern. Many now agree that attention should move away from policy production to the implementation of existing policies to forestall any further loss in the EU's legitimacy and effectiveness.[19]

In order to address the complexity of the EU policy process and particularly the problems of EU policy implementation, analysts have used various analytical approaches. Traditional *intergovernmentalists* have studied EU politics from a state-centrist perspective and have focused on national actors and their influence on the EU policy process.[20] Other analysts have resolved to approaches which accommodate more appropriately the increasing complexity of the EU: *policy networks* analysts have tended to ignore governmental levels and have focused instead on actors and their interests during the bargaining 'game', while *multi-level governance* analysts have focused on essentially three government levels and their influence on EU politics.

Policy Networks

Among the pioneers of *policy networks* analysis have been Rhodes and Marsh (1992) who put some order into the complexity of UK and EU policy processes by identifying a continuum of *policy network* types ranging from *policy community/ territorial communities* to *issue networks*.[21] To help identify the type of network, Rhodes and Marsh suggest three broad criteria:

- the relative stability and continuity of network membership
- resource dependencies (resources can be of constitutional-legal, organisational, financial, political or informative nature)
- the relative insularity and autonomy of a network from outside influences (i.e. other networks).

Policy networks identify actors, their interests, positions and resources in the policy process. Moreover, *policy networks* highlight the complexity of modern politics characterised by a decline in governmental control and an increasing dominance of coalitions consisting of governmental as well as non-governmental actors sharing certain interests. These new coalitions tend to overstep conventional horizontal and vertical boundaries and create new policy process constellations. *Policy networks* can be applied to the EU context: they accommodate the complex patterns of EU interdependencies and enmeshing of interests, as well as the emergence of new values and coalitions in the EU. In recent years, analysts such as Heritier (1993 and 1994); Bomberg (1994); Bressers et al (1994); McAteer and Mitchell (1996) have discovered the usefulness of *policy networks* for their studies of EU politics and policy-making.

At first glance, *policy networks* constitute a powerful analytical tool. There are caveats, however. In general, there is a danger of *policy networks* developing into 'giant garbage can[s]',[22] containing all possible aspects but providing no basis for a clear and structured analysis. Another problem concerns the distinction between (and definition of) *policy network* categories: the line between networks cannot always be drawn, as many of them are inter-connected or interdependent. As far as implementation studies are concerned, *policy networks* are useful in so far as they highlight differences between policy-making elites and established network coalitions on the ground which may oppose and resist new policies.[23] However, while *policy networks* by definition describe actors, their positions, resources and interests, they downplay formal institutional factors such as statute forms and constitutional structures and neglect other circumstances such as events and geographical conditions. *Policy networks* therefore do not constitute an analytical tool which covers all major aspects of EU policy implementation.

Multi-level Governance

Another analytical approach, promoted by EU analysts such as Marks et al (1996); Hooghe (1996); Jeffery (1996, 1996a and 1996b); and Scharpf

(1994) responds to the increasing pressures of *sub*national actors who have sought to influence the EU policy process more forcefully. Their approach offsets the state-centrist perspective[24] by attributing weight to essentially three government levels in the EU policy process: the EU, the national, and the *sub*national levels. Many of the *multi-level governance* studies have focused on the interactions across government levels during the bargaining process and each government level's influence on EU policy decisions. As far as the latter stages of the EU policy process are concerned, *multi-level governance* studies exist, however they concern predominantly experiences in the EU Structural Policy,[25] a policy area that differs in many respects from the EU environmental policy area. In essence, the EU Structural Policy involves the allocation of funds to support disadvantaged regions and social groups. One of its key principles is the 'partnership' between EU, national and *sub*national actors during both policy-making and implementation (i.e. during the negotiation and allocation of financial support and the subsequent pursuance of regional and social projects). Since the Structural Funds regulations explicitly mention (and are designed to suit) the 'partnership' of actors at all government levels, the EU Structural Policy is a convenient study area for *multi-level governance* analysts. The involvement of essentially three government levels is less obvious in the EU environmental policy area as *sub*national authorities' responsibilities are not outlined explicitly in EU environmental Directives. However, this implicit 'partnership' does not mean that the *multi-level governance* approach cannot be applied to the study of EU environmental policies and their implementation. While the process of EU environmental policies is more complex and more difficult to trace than the EU Structural Policy process, the *multi-level governance* approach still promises to be a useful tool for this research area.

EU Implementation Studies

A number of EU studies have focused on implementation experiences and have generated some valuable findings. However, many EU implementation studies have shown limitations: some have tended to be descriptive and failed to provide a conceptual basis for further research, while others have not followed a systematic comparative analysis which could have highlighted more forcefully key problems of EU policy implementation.

The first study is interesting considering the time of publication: in 1975 Puchala presented his findings on *EC post-decisional politics* which have since been re-confirmed by more recent implementation studies. Puchala emphasised the need to study the whole policy process which includes the implementation of EU policies at lower government levels. According to Puchala, EU politics involves the harmonisation of Member States' national policies so that a common, mainly economic, ground can be established. This harmonisation process inevitably implies changes which may benefit some but harm others at the 'domestic' level. Member States' governments usually find themselves caught between the European Commission's call for compliance of EU policy commitments and resistance from various interest groups within the Member States. This conflictual situation is even more problematic when a Member State's government is divided over one particular policy or has accepted reluctantly an EC policy in return for other benefits. In the latter case, a Member State may exercise a 'second veto' at the later, post-decisional stage when an EU policy is supposed to be implemented.[26] Faced with pressure from above (EU) and below (domestic constituencies), Member States' governments tend to give way to domestic pressures. However, since most EU policies are legally binding, Member States will eventually follow their obligations. The European Commission recognises the dilemma faced by Member State governments and is aware of its own limited enforcement powers. It therefore takes a pragmatic position and is careful not to demand the 'impossible' from Member States.

With his study on *EC post-decisional politics*, Puchala made a valuable contribution to the research of EU policy implementation. He pointed out the unique pressures and dilemmas associated with EU policies and the differences of interests and motivations depending on government levels and policy stages. A policy idea, feasible and attractive at the EU policy-making stage, may be unfeasible and unwelcome at the practical implementation level. In particular, EU environmental policies adopted at the EU level (for environmental and often economic 'level-playing-field' reasons) almost always involve costs at a later stage and are often perceived as unbearable burdens by domestic implementors. The discrepancy between EU policy objectives and domestic interests goes some way towards explaining the 'implementation gap'. Following Puchala's logic, the wider the gap between EU and domestic interests, the less likely it is that an EU policy is implemented properly.

However, Puchala's study represents a state-centrist account of the EU post-decisional phase. It focuses on national governments and their 'sandwiched' position between EU policy obligations and domestic pressures as a whole. Puchala neglects variations of domestic pressures treating them as homogenous entities. Moreover, he is only interested in domestic pressures as they affect Member States' bargaining at the EU level. Puchala's domestic perspective therefore needs to be supplemented for this research. In fact, it could be argued that Puchala's approach does not provide for an accurate picture of EU (environmental) policy implementation because it runs the risk of ignoring those *sub*national variances which significantly shape EU policy implementation.

Following a different approach, Siedentopf and Ziller edited a two-volume study in 1988 which lists the (then) twelve EC Member States and their implementation performances of 17 EC Directives.[27] Their twelve case studies represent the first comprehensive attempt to describe, compare and assess EU policy implementation results and problems. However, the authors of the case studies did not conduct their research in a comparable pattern and their investigative structures and approaches differed depending on their personal preferences. Nevertheless each contribution provides insights into the practice of EU policy implementation. For instance, in their case study on Germany, Pag and Wessels include an EU-specific criterion which cannot be found in other policy systems. The criterion *general integration attitudes* refers to implementors' attitudes towards the EC/EU which can vary over time depending on political and economic circumstances. Pag and Wessels' criterion is important because favourable attitudes towards the EU can facilitate the implementation of EU policies, while critical attitudes can hinder the implementation of EU policies. In either case, *EU integration attitudes* influence the process of EU environmental policy implementation.

From a more legalistic perspective, Krämer (1991, 1992 and 1997) highlighted the gap between EU environmental legislation and Member States' non-compliance which he considered as more alarming in the environmental policy area than in any other policy area. Krämer did not formulate an analytical concept that explains the alarming gap. Instead, he identified areas of non-compliance, differences in formal and practical implementation among the Member States, and their limited commitment in EU environmental policy obligations. Krämer did not establish a conceptual framework but his first-hand insights nevertheless stress the importance to further investigate (and solve) the problems of EU environmental policy implementation.

From a more comparative angle, Butt Philip (1994) investigated the EU environmental and social policy areas. In comparison with the EU social policy area, Butt Philip pointed out that the increase in complaints concerning environmental policy non-compliance had been 'spectacular'. He blamed inadequate consultation, deliberate ambiguities in legal texts, inconsistencies in policy objectives, lack in administrative and financial resources, and the ineffectiveness of existing penalties for inadequate policy implementation. Similar to Krämer's studies, Butt Philip's comparison emphasises that the implementation (and compliance) of EU environmental policies is more complex and problematic than other EU policy areas and therefore deserves particular research attention.

Finally, Collins and Earnshaw (1992) presented the flaws in EU environmental policy implementation across Member States and across three broad implementation stages.[28] They identified five main reasons for the Member States' poor environmental policy implementation results:

- the complexity of the transposition process from the EC level to the national level
- misinterpretations of legal texts
- structural obstacles
- legislative cultures which may be incompatible with EU legislation
- political considerations which may have a hindering impact on the EU legislation.

Again, Collins and Earnshaw did not establish a conceptual basis tailored for this research area. Instead their investigation relies upon the analytical approaches of other analysts, in particular Pressman and Wildavsky.

All of the above studies are relevant and contribute to a better understanding of the processes and problems of EU environmental policy implementation. However, they do not fully cover and conceptualise the research matter. Referring to the great difficulties in grasping the EU and its policy process, Schumann (1991) commented that 'it is necessary to embrace the whole elephant'. In the EU environmental policy area, the beast appears to be even more difficult to embrace. The following outline is an attempt to do just that by providing an analytical framework which synthesises and enhances existing studies.

Drawing-up a Multi-Layered Implementation Map

The following *multi-layered implementation map* captures the complex EU environmental policy process.

Figure 2.1: The Multi-Layered Implementation Map

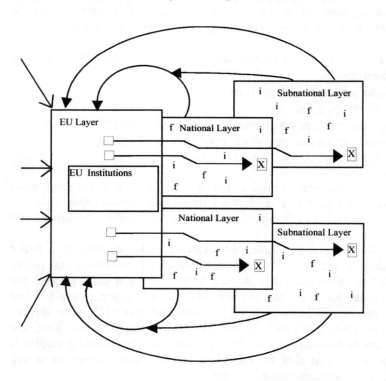

Key:

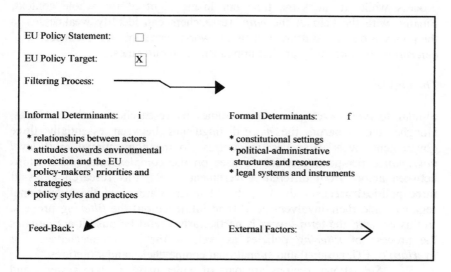

In essence, the *map* describes the implementation of EU environmental policies as a *filtering* process. Policies are not translated directly but *filtered* through unique political systems, or *layers*. In contrast to other *multi-level governance* studies, which focus on the inter-active dynamism between government levels and their actors, this *map* highlights and distinguishes between three government *layers* - the EU, national and *sub*national *layers*. The *layers* feature their own mixes of formal and informal *determinants* which either facilitate or hinder EU environmental policies reaching their implementation *targets*.

The *map* fulfils essentially three purposes:

- it guides the investigation through the EU environmental policy process and provides a comparative framework
- it accommodates relevant aspects of the process and highlights key *determinants*
- it provides a template for further investigations of EU policy implementation problems.

The *map* does not suggest a new challenging theory for the discipline which predicts future policy outcomes. Rather, by way of focusing on government levels and identifying and sorting implementation *determinants*, the *map* captures the details of the EU environmental policy process while at the same time not losing sight of the whole context. Finally, with the help of the *map*, researchers can identify weaknesses in the policy process and draw-up lists of 'worst scenario' and 'ideal scenario' *determinants* which help predict implementation outcomes.

The Layers

Similar to *multi-level governance* studies by researchers such as Marks, Hooghe and Scharpf, the *map* distinguishes between essentially three government levels. However, in contrast to the conventional *multi-level governance* perspective which focuses on the complex inter-relationships between actors from different government levels, this research looks at three political arenas - the EU, the Member States, and the *sub*national regions - and their involvement in (and influence on) the *filtering* process. In this context, the term *layers* is particularly useful because it accentuates the process of *filtering* policies as well as the inter-connectedness (or *overlap*) of EU, national and *sub*national competencies and structures.[29]

    *Sub*national regions are part of wider national state systems and are influenced by national conditions. Yet, while both national and *sub*national *layers* are to a certain extent inter-connected and feature some similarities in the implementation of EU environmental policies, they also feature different conditions which shape EU environmental policies accordingly. The distinction between *layers* is therefore important for a refined study of EU environmental policy implementation which generally involves actors at both national and *sub*national levels. In this context, the *map* clearly departs from Puchala's domestic variable by distinguishing between national and *sub*national *layers* and investigating both *layers* separately. Studying domestic conditions as a homogeneous whole neglects and underestimates those *sub*national *determinants* (geographical, cultural, constitutional etc.) which have an impact on EU environmental policy outcomes. Moreover, *sub*national actors process EU environmental policies within their own systems and in their own way, which may well differ from national practices. The domestic perspective would therefore not provide an adequate picture of the EU environmental policy 'reality'.

    Strictly speaking, there are government levels below the *sub*national level (i.e. regional, local levels). The composition and structures of government levels inside the *sub*national *layers* depend upon the regions themselves and their constitutional settings. Subdividing the *sub*national *layer* into further *layers* would refine even more the research picture of EU environmental policy implementation. However, the purpose of this research is to highlight the importance to move 'beyond' the national (or domestic) level when investigating EU environmental policy implementation and to provide a template for further investigation. Expanding the *multi-layered implementation map* to add more *layers* to the investigation was therefore considered unnecessary.

The Arrows

EU environmental policies are not static phenomena: they are developed, negotiated and adopted, and later require implementation (and enforcement) at either the national level or *sub*national level. Focusing on policy-making factors, the *map* distinguishes between *external factor arrows* and *feed back arrows*. Firstly, EU environmental policy-making is influenced by *external factors* in the form of international pressures (for instance, commitments to UNCED, global environmental problems, and international economic agreements) and pressing environmental issues such as pollution incidents and problems of long-term environmental deterioration. These factors occur sporadically and can have an indirect impact on EU environmental policy-making. EU environmental policy-making is influenced more frequently by the *feed-back* from actors at the national and *sub*national government levels. From the national level central government ministers (in the Council of Ministers), COREPER officials, national experts and advisors as well as 'nation-wide' interest groups seek to influence EU environmental policy-making. In addition, *sub*national politicians, administrators, experts and regional or local interest groups seek to channel their views on EU environmental policies to the national and EU levels.[30] EU institutions themselves (i.e. Council, EP, Commission, EEA, EcoSoc, CoR, to a certain extent ECJ) contribute to the form and content of EU environmental policies. All these forces (some are more influential than others) contribute towards a complex, and often cumbersome, EU policy-making process and influence the form and content of the policies themselves.[31]

It is important to consider the complexity of the early stages of the EU environmental policy process when studying policy implementation and policy outcomes. Conditions which contribute towards the adoption of EU environmental policies vary: the Council in particular, adopts policies either because Council ministers support the policy objectives, because ministers accept them in order to pursue other policy goals, or because they hope to ignore them at the later implementation stage (Puchala termed the latter option 'second veto'). Implementors observe the process of policy adoption and act accordingly; their commitment is influenced later by either the policy-makers' whole-hearted support or reluctance towards a policy. Whether or not their opinions and lessons from past implementation experiences are taken into account at a later policy-making stage (indicated with *feed-back* arrows) is also an important aspect for implementors and their commitment towards EU environmental policy objectives. The *map* acknowledges the complex relationship between EU policy-making factors and the dynamism of EU bargaining on the one hand and EU policy implementation circumstances on the other.[32]

Focusing on the latter stages of the process, the map describes the implementation of EU environmental policies as a *filtering* process through the national and *sub*national *layers*. The majority of EU environmental policies concern areas which predominantly touch upon *sub*national competencies (for instance planning, water and waste management) and therefore require implementation and compliance at the *sub*national level. Other EU environmental policies concern the national level only. The *map* therefore distinguishes between *filtering arrows* pointing towards national and *sub*national *layers*. Since there are more EU environmental policies that require implementation on the *sub*national (and even local) ground, *filtering arrows* pointing towards *sub*national *layers* should outnumber the national *filtering arrows*. For the sake of simplicity, however, each implementation process is indicated in the *map* with one *arrow* only.

In order to reach the *sub*national ground, EU environmental policies generally have to be *filtered* through the national *layer* which, to a certain extent, pre-shapes the policies. However, most of the formal transposition (i.e. adjustments of the legal framework to accommodate EU Regulations and Directives) and practical implementation (i.e. application in practice of EU policies' standards and objectives) are conducted within the *sub*national *layer*. It is therefore important to study the influence of the national *layer* but pay particular attention to the *sub*national *layer* and its influence on EU environmental policy outcomes.

Policy Statements and their Targets

As is typical for a more traditional implementation study, the *map* highlights *policy statements* and their *targets*. In contrast to mere statements of political opinions (or *standpoints*, see Pressman and Wildavsky), EU environmental policy statements combine and accommodate a wide range of interests and objectives which require realisation at a later stage. EU environmental *policy statements* (i.e. Regulations and Directives) are legally binding and should therefore be implemented by the Member States and their *sub*national regions. However, while Directives (constituting the majority of EU environmental policies) outline legally binding policy objectives, they generally leave the ways and means to the Member States and their national and *sub*national implementors. Directives are therefore comparable with the above-described *Wilsonian* statutes which involve a complex sharing of discretionary powers between policy-makers and implementors. This sharing of powers often results in policy outcomes which do not resemble the original policy objectives as outlined in the statutes (see X in *map*).

Given the complexity of powers and interests, EU environmental policy objectives or *targets* (X) are either reached, are missed by national and *sub*national implementors, or EU environmental policies result in unexpected and unintended policy outcomes.[33] In the light of three government *layers* involved in the EU environmental policy process and the potential obstacles for environmental policy objectives from other policy areas and interests, the majority of EU environmental policies are most likely to 'get stuck' in the *filtering* process falling short of the policy *target*. Or EU environmental policies take shape different from their objectives stated on paper. In the latter case, EU environmental policies fall short of *target* as well.

Formal and Informal Determinants

The *map* identifies *determinants* which shape EU environmental policy outcomes.[34] Ultimately, these *determinants* can facilitate, divert or prevent EU environmental policies in reaching their *targets*. The *map* distinguishes between *formal* and *informal determinants* which can be found in 'unique' combinations in each *layer*. In order to gain a structured overview, the *map* isolates seven categories of *determinants*:

Formal determinants
- constitutional settings
- political-administrative structures and resources
- legal systems and instruments.

Informal determinants
- relationships between actors
- attitudes towards environmental protection and the EU
- policy-makers' priorities and strategies
- policy styles and practices.

The *determinant* categories best encompass the research matter: they take account of the different government levels (or political arenas) involved in the EU environmental policy process and accommodate the complex environmental policy dimension which tends to involve a multitude of policy areas, interests and competencies. The *determinants* also allow the researcher to investigate EU environmental policy implementation from a micro-perspective (for instance, the researcher can focus on the details of formal transposition of EU environmental policies into national and *sub*national legal systems), while assessing the overall and combined impact of all *determinants* on EU environmental policy implementation from a macro-level.

When EU environmental policies are *filtered* through the *layers*, they are either blocked by *formal* and *informal determinants*, facilitated by favourable *determinants*, or their content and objectives are altered by *determinants* on the ground (see Argument 1). Similar to existing 'ideal implementation scenario' check lists (see in particular Sabatier and Mazmanian), the *map* can be used to assess whether *determinants* in the national and *sub*national *layers* are favourable or unfavourable. Again, by following the *map*, the researcher can investigate details of the implementation process while taking account of the process as a whole.

Considering the multitude and complexity of government *layers* and the complexity of environmental issues and their policy solutions, *determinants* which influence EU environmental policies during the *filtering* process are expected to be more diverse and numerous than in other policy areas. Certain *formal* and *informal determinants* stand out in the EU environmental policy process. In the *formal determinant* category, the compatibility of EU legal instruments vis-à-vis national and *sub*national legal systems and administrative structures plays a significant role in the success of EU environmental objectives. If the form or content of an EU legislation is incompatible with national and *sub*national structures and legal frameworks, the transposition of EU environmental policy objectives is likely to fail. In turn, if the form or content of an EU policy is compatible with national and *sub*national formal conditions, the EU environmental policy is likely to succeed unless there are major informal obstacles in the form of hostile relationships between actors[35] and conflicting policy priorities which prevent the policy from being implemented.

Focusing on *informal determinants*, both attitudes towards the EU and the relationships between EU and national/ *sub*national actors can have an impact on EU environmental policy implementation. Past implementation studies (such as Pag and Wessels' case study mentioned above) have shown that among other factors, the success of EU policies has depended very much upon implementors' attitudes towards the EU and its 'legitimacy' to produce common policies. National and *sub*national implementors have also been influenced by their formal links and informal relationships with EU actors. Both aspects should therefore be taken into account when exploring EU environmental policy implementation.

Another important *informal determinant* concerns the prioritisation of economic interests. Economic considerations still play a significant role for actors during environmental policy processes (see Offe's argument). In the context of EU environmental policies, the economic imperative is particularly relevant. During EU environmental policy-making, economic considerations either hinder the adoption of 'costly' environmental policies, or they motivate the adoption of harmonised environmental standards for the sake of an economic 'level-playing-field'. At the implementation stage, economic considerations take a different shape: in many cases, the economic motivation to harmonise environmental standards in the EU diminishes in the light of implementors' (self-) interest to protect national and *sub*national economies. Unless EU environmental policies are convenient and complementary to economic priorities within the national and *sub*national *layers*, economic considerations can become obstacles in the implementation path of EU environmental objectives. In any case, economic considerations play a vital role in the pursuance of EU environmental policies. They either complement and support environmental objectives or do not touch upon environmental objectives. In both cases EU environmental policies can be implemented without problems. Economic considerations can also clash with environmental objectives. In the latter case, EU environmental policies are likely to fail on the ground. By and large, economic incentives and 'trade-offs' between EU policy-makers (especially in the Council of Ministers) which result in the adoption of many EU environmental policies are almost absent in the minds of implementors at a later stage with the result that EU environmental policies often do not receive the necessary backing on the ground.

Comparing Implementation Experiences

As Lipset (1994) once commented,

a person who knows only one country basically knows no country well.

The same rule applies to the study of EU Member States and their *sub*national regions. The *map* provides for a systematic comparison of implementation performances at both national (UK and Germany) and *sub*national (Scotland and Bavaria) levels. A thorough investigation and comparison of both national and *sub*national *layers* is necessary in order to gain a comprehensive picture on the overall process of EU environmental policies. The *map* guides the reader: it helps establish the differences and similarities between the UK and Germany first before attending to the similarities and differences between Scotland and Bavaria. It also compares the *sub*national regions with their 'mother' states. This two-dimensional assessment not only promises to uncover a wide range of *determinants* which decisively shape the implementation of EU environmental policies, it also contrasts divergencies between the *layers* under investigation.

Possible Caveats and Limitations of the Map

The *map* does not establish a new theory or blueprint for predicting future policy outcomes. It does not establish a causal link between policy *statement* and *target*, neither does it predict and quantify the effectiveness of EU environmental policies. The *map* is a guiding tool which helps investigate the complexity of the *filtering* process and identifies obstacles which hinder the implementation of EU environmental policies in the *layers*. Similarly, with the help of the *determinant* categories, researchers can draw-up a list of 'ideal scenario' factors which facilitate the implementation of EU environmental policies.

The *map* resembles what some researchers would call a traditional *top-down* approach towards EU environmental policy implementation whereby political outcomes different from the original policy objective are seen as negative outcomes or failures. True, this research investigates critically the shortcomings of EU environmental policy implementation and highlights problems and obstacles during the process. Yet, it does not point the finger at one particular government level or one particular group of actors. Instead, it aims to contribute towards a better understanding of the whole EU environmental policy process and the problems associated with implementation. While the *map* resembles a *top-down* investigation, it nevertheless makes some allowance for a more dynamic policy approach by including *feed-back arrows* and other (external) *arrows* which influence EU environmental policy as a whole.

There are some aspects which the *map* cannot possibly include without causing confusion. Firstly, the *map* does not describe in detail the policy-making procedures that take place within the *layers*. The *map* is intended to focus mainly on the latter stages of the policy process (i.e. policy implementation). It therefore includes only policy-making *arrows* (i.e. *external factors* and *feed-back arrows*) which help explain the style and content of EU environmental policies as well as their subsequent implementation. Similarly, the *map* does not highlight the Commission as a typical *policy broker* (see Sabatier) and representative of the *political-administrative subsystem* (see Offe) during policy formulation. The Commission plays a vital role during policy-making process. However, when it comes to the implementation of most EU environmental policies, the Commission plays a predominantly guardian function. It is therefore sufficient to recognise indirectly the role of the Commission as part of the EU institutions (indicated inside the EU *layer*) and as a facilitator (or obstacle) when investigating the *filtering* process. The ECJ, too, is not highlighted in the *map* although it plays a central role in the interpretation and enforcement of EU environmental legislation. However, in the context of *filtering* EU environmental policies, it is not directly involved in the actual process. It is therefore sufficient to acknowledge the ECJ as part of the EU institutions box and refer to it whenever implementors in the national and *sub*national *layers* are affected by its judgments.

The *map* does not distinguish between different types of EU legal instruments or policy *statements*. While the form and content of legal instruments constitute important factors in the implementation of policy objectives (see Berman, Ingram and Schneider), the instruments' categorisation would make the *map* less intelligible. Since the main focus of this research is the *filtering* process through government *layers*, it was decided not to include statute categories in the *map*.

Finally, the *layers* themselves are featured in the *map* as identical in terms of shape and size. This does not reflect the differences and similarities of the two Member States and their *sub*national regions under investigation. Indeed, focusing on the *sub*national regions, Bavaria and Scotland differ in many respects. They provide a stark contrast in terms of their embeddedness in two opposite state systems: Scotland was until recently part of a centralised state system and has undergone a devolution process, while Bavaria has been, and still is, part of a federal state system. Bavaria and Scotland also feature differences in the way their political-administrative actors and citizens perceive (and deal with) environmental matters and policies. In addition, Scotland and Bavaria differ in geographical terms: Scotland represents a peripheral region of the EU with its own unique characteristics and problems (such as difficulties in accessing EU markets), Bavaria is situated more or less in the centre of Europe coping with economic and other pressures from Central and Eastern Europe. On the other hand, Scotland and Bavaria feature certain similarities such as a strong territorial identity, relatively large natural resources (which are of EU significance), as well as expanding high technology industries. Obviously, the *map's* uniform *layers* do not describe the *sub*national regions fully. However, for the purpose of comparative analysis (which can be applied to other EU *sub*national regions), it is sufficient to arrange the *layers* in a simplified and systematic manner.

On a similar note, the regions feature government levels below which are structured differently depending on their constitutional and political-administrative settings. The *map* could therefore include more *layers* to take account of the variances inside the *sub*national regions. However, it would be difficult to determine exactly how many *sub-subnational* layers should be added to the *map*. Moreover, the main purpose of this research is to move beyond the national government level when investigating EU environmental policy implementation. The *multi-layered implementation map* as it stands should therefore provide adequate guidance for the researcher.

The above synthesis of analytical approaches has already suggested that the scenario for the implementation of EU environmental policies is far from ideal. The *map* seeks to combine all relevant aspects that shape EU environmental policy implementation into one comprehensive framework. Its strength lies in its potential to guide investigations on EU environmental policy implementation at both macro and micro levels. It distinguishes between national and *sub*national arenas (i.e. *layers*) and puts order into the complex mix of formal and informal factors (i.e. *determinants*) that shape EU environmental policies in the process. With the help of the *map*, the following Chapters will address the implementation deficit in the EU environmental policy area by examining step by step the EU, the national and the *sub*national *layers*. The Chapters will identify and highlight the *layers'* key *determinants* and will assess to what extent national and particularly *sub*national *layers* influence the success, or failure, of EU environmental policies.

Notes

[1] See, for instance, Collins and Earnshaw (1992) and Sabatier and Mazmanian (1980).
[2] This Program was produced as an employment creation measure in a predominantly black

community facing high unemployment and associated problems such as crime and poverty.

3   O'Toole (1988); Pag and Wessels (1988) argue that the number of actors is insignificant if a policy is generally perceived as positive and there are no obvious administrative or political obstacles. They prefer a combination of quantitative and structural-compositional criteria as a measurement.

4   Meny et al (1996) describe the 'push-effect' of the Netherlands, Denmark and Germany on EU environmental policy-making.

5   The term of 'lowest common denominator' is used by Moravscik (1991).

6   For a critique see in particular Jordan (1996).

7   In the 1990s the European Commission adjusted its environmental policy strategy to allow for implementors' views and problems. For further details see Chapter 3.

8   See, for instance, Comfort (1980).

9   See O'Brien (1980).

10 For Thompson's typology of interdependence see O'Toole and Montjoy (1984).

11 A *pooled* relationship between policy-makers and implementors implies a simple, straightforward policy instruction; the *sequential* category involves intermediate steps which can have a hindering impact on the policy implementation; *reciprocal* interdependence refers to a more complex relationship between policy-makers and implementors which requires more coordination at all stages of the policy process.

12 The *strong statute* represents a detailed piece of legislation with precise instructions from policy-makers to implementors; the *grass roots statute* constitutes a mere encouragement from above to shape and implement a policy at the lowest possible government level; the *support building statute* tends to be a complementary piece of legislation which relies upon voluntary policy implementation and compliance; the *Wilsonian statute* can be located between the two extremes of the *strong* and *grass roots statutes*: the *Wilsonian Statute* is strong in goal specificity but leaves wide discretion to implementors on the ways and means in achieving policy goals.

13 'A directive shall be binding, as to the result to be achieved, upon each Member State to which it is addressed, but shall leave to the national authorities the choice of form and methods.' (Art.189 EC, new: Art.249EC)

14 While Sabatier emphasises the process of learning which results from the exchange of views, Offe highlights the tensions between the subsystems.

15 The Commission pushed hard for Council Regulations (EEC) 880/92 and 1836/93 introducing 'eco-labelling' and 'eco-management' schemes, based initially on voluntary participation of the industrial sector and promoting sustainable products and production as a business opportunity.

16 DG XI responsible for environmental matters has to co-ordinate policy objectives with other DGs such as DG VII (transport) and DG XVI (regional policy).

17 One such initiative is the EU Commission's Task Force which produced a *Report* (1990) on the environment and the Internal Market.

18 Ken Collins (until 1999 MEP and chairman of EP Environment Committee) used the term 'paradox' during an interview, 3. February 1995, East Kilbride. Jänicke et al (1996) provide empirical evidence for the 'paradox'. Grossman (1995) discusses the complex relationship between environmental and economic matters.

19 In a different context (i.e. in the context of the EU Cohesion Policy), Hooghe (1998) points out that the EU's legitimacy is currently on an insecure footing. EU policy-makers find it increasingly difficult to justify their capacity to adopt legislation which is

later not implemented properly. Hooghe describes this dilemma as policy dysfunctionality.

[20] See in particular Moravscik (1993).

[21] Apart from Rhodes and Marsh (1992) see also Rhodes (1988 and 1997).

[22] Quotation from Bressers et al (1994).

[23] Rhodes and Marsh (1992a) conclude that many 'Thatcherite' policies failed because of the continued importance of *policy networks* on the ground.

[24] For an example of a state-centrist analysis see Moravscik (1993). For a discussion of state-centrist and *multi-level governance* perspectives see Marks, Hooghe and Blank (1996).

[25] See, for instance, Marks (1993).

[26] Lenearts (1994) even suggests that in the past 'Member States did not use the right of veto against legislation whose implementation would in any event be weak and lightly monitored'.

[27] The 17 Directives selected for the case studies ranged from an Electrical Equipment Directive (76/117/EEC) to a Company Accounts Directive (78/660/EEC).

[28] Implementation stages: First - transposition of EU legislation into national law; second - practical results and impacts; third - enforcement and monitoring of legislative obligations.

[29] Concentric circles were considered as an alternative to the term *layers*. However, since the central theme of this research is the *filtering* process and the distinction between national and *sub*national implementation performances, overlapping *layers* are more suitable as an illustration of government levels. In the same context, the author prefers the *multi-level governance* description of political arenas being 'inter-connected' rather than 'nested'. Marks et al (1996) make a similar distinction.

[30] Depending on the Member State and depending on the policy issue, some *sub*national representatives can participate in the Council of Ministers.

[31] For detailed accounts of EU policy-making see, for instance, Nugent (1999), Dinan (1999) and Bomberg and Peterson (1999).

[32] The dynamic relationship between government levels and policy processes is described by Weale (1996) as a system of European governance.

[33] The *targets* resemble Pressman and Wildavsky's *end points* which have to be reached in order for a policy to be effective.

[34] The author considered other terms such as *attributes* or *variables* as alternatives to *determinants* but decided that *determinants* best capture the intended description of key factors which shape EU environmental policy implementation one way or another.

[35] Thompson (see above) argues that relationships between policy-makers and implementors play a decisive role in the success of policies.

References

Berman, P. (1980), 'Thinking about Programmed and Adaptive Implementation: Matching Strategies to Situations', in Ingram, H. and Mann, D.E. (eds) *Why Policies Succeed or Fail*, Sage, London.

Bomberg, E. (1994), 'Protecting the Periphery: Environmental Policy in Peripheral Regions of the European Union', *Regional Politics & Policy*, vol.4, No.1, pp.45-61.

Bressers, H. et al (1994), 'Networks as Models of Analysis: Water Policy in Comparative

Perspective' *Environmental Politics*, vol.3, No.4.

Butt Philip, A. (1994), *Regulating the Single Market: A Comparison of the Implementation of Social and Environmental Legislation*, Research Paper.

Collins, K. and Earnshaw, D. (1992), 'The Implementation and Enforcement of EC Environment Legislation', *Environmental Politics*, vol.1, No.4, Winter, pp.213-249.

Comfort, L.K. (1980), 'Evaluation as an Instrument for Educational Change', in Ingram, H. and Mann, D. *Why Policies Succeed or Fail*, Sage, London.

Commission (1990), '*1992' The Environmental Dimension. Task Force Report on the Environment and the Internal Market*, Report.

Dinan, D. (1999), *Ever Closer Union? An Introduction to the European Community*, MacMillan, Basingstoke.

Elmore, R.F. (1979/1980) 'Backward Mapping. Implementation Research and Policy Decisions', *Political Science Quarterly*, vol.94, pp.601-616.

Grossman, G.M. (1995), 'Pollution and growth: what do we know?', in Goldin, I. and Winters, L.A. (eds) *The Economics of Sustainable Development*, Cambridge University Press, Cambridge.

Heritier, A. (1993), 'Policy-Netzwerkanalyse als Untersuchungsinstrument im europäischen Kontext', *Politische Vierteljahresschrift*, Sonderheft: Policy-Analyse. Kritik und Neuorientierung, 24/93, pp.432-449.

Heritier, A. et al (1994), *Die Veränderung von Staatlichkeit in Europa. Ein regulativer Wettbewerb. Deutschland, Großbritannien, Frankreich*, Leske + Budrich.

Hill, J.S. and Weissert, C.S. (1995), 'Implementation and the Irony of Delegation: The Politics of Low-level Radioactive Waste Disposal', *The Journal of Politics* vol.57, No.2, May, pp.344-366.

Hooghe, L. (ed) (1996), *Cohesion Policy and European Integration. Building Multi-level Governance*, Clarendon Press, Oxford.

Hooghe, L. (1998), 'EU Cohesion Policy and Competing Models of European Capitalism', *Journal of Common Market Studies*, vol.36, No.4, December, pp.457-477.

Ingram, H. and Schneider, A. (1990), 'Improving Implementation through framing smarter Statutes', *Journal of Public Policy*, vol.10, pp.67-88.

Jänicke, M. et al (1996), *Umweltpolitik der Industrieländer*, edition sigma, Berlin.

Jeffery, C. (1996), *The Emergence of Multi-level Governance in the European Union*, Conference Paper, 8eme Colloque International de la Revue 'Politique et Management Public'.

Jeffery, C. (1996a), 'Sub-National Authorities and European Domestic Policy', *Regional and Federal Studies*, vol.7, No.3, pp.204-219.

Jeffery, C. (1996b), 'Towards a Third Level in Europe? The German Länder in the European Union', *Political Studies*, vol.44, No.2, pp.253-266.

Jordan, A. (1996), 'The European Union: Rigid or Flexible Decision-Making. From Brussels to Blackpool and Southport. Post-decisional Politics in the European Community', *Contemporary Political Studies*, Hampster-Monk (ed), PSA Conference Proceeding, pp.1897-1906.

Jordan, A. (1996a), *Implementation Failure or Policy Making? How Do We Theorise the Implementation of EC Policy at the National and Sub-national Level?*, Working Paper, Centre for Social and Economic Research on the Global Environment, University of East Anglia and University College of London.

Krämer, L. (1991), 'The Implementation of Community Environmental Directives within Member States: Some Implications of the Direct Effect Doctrine', *Journal of*

Environmental Law, vol.3, No.1, pp.39-56.

Krämer, L. (1992 and 1997), *Focus on European Environmental Law*, Sweet and Maxwell, London.

Lenearts, K. (1994), 'The Principle of Subsidiarity and the Environment in the EU. Keeping the Balance of Federalism', *Fordham International Law Journal*, vol.17, Part 4, pp.846-895.

Lipset, S.M. (1994), 'Binary Comparisons. American Exceptionalism - Japanese Uniqueness', in Dogan, M. and Kazancigil, A. (eds) *Comparing Nations. Concepts, Strategies, Substance*, Blackwell, Oxford.

Marks, G. (1993), 'Structural Policy and Multilevel Governance', in Cafruny, A. and Rosenthal, G. (eds) *The State of the European Community. The Maastricht Debates and Beyond*, vol.2, Lynne Riemer.

Marks, G. et al (1996), *Governance in the European Union*, Sage, London.

Marks, G., Hooghe, L. and Blank, K. (1996), 'European Integration from the 1980s: State-Centric v. Multi-level Governance', *Journal of Common Market Studies*, vol.34, No.3, September, pp.341-378.

McAteer, M. and Mitchell, D. (1996), 'Peripheral Lobbying! The Territorial Dimension of Euro Lobbying by Scottish and Welsh Sub-central Government', *Regional and Federal Studies*, vol.6, No.3, pp.1-27.

Meny, Y. et al (1996), *Adjusting to Europe. The Impact of the European Union on National Institutions and Policies*, Routledge, London.

Moravscik, A. (1991), 'Negotiating the Single European Act: National interests and conventional statecraft in the European Community', *International Organisation*, vol.45, pp.19-56.

Moravscik, A. (1993), 'Preferences and Power in the European Community: A Liberal Intergovernmental Approach', *Journal of Common Market Studies*, vol.31, No.4, December, pp.473-524.

Nugent, N. (1999), *The Government and Politics of the European Union*, MacMillan, Basingstoke.

O'Brien, D. (1980), 'Crosscutting policies, uncertain compliance, and why policies often cannot succeed or fail', in Ingram, H. and Mann, D. *Why Policies Succeed or Fail*, Sage, London.

Offe, C. (1984), *Contradictions of the Welfare State*, Hutchinson, London.

O'Toole, L. (1986), 'Policy Recommendations for Multi-Actor Implementation. An Assessment of the Field', *Journal of Public Policy*, vol.6, part 2, pp.181-210.

O'Toole, L. (1988), 'Strategies for Intergovernmental Management: Implementing Programs in Interorganizational Networks', *International Journal of Public Administration*, vol.11, pp.417-441.

O'Toole, L. and Montjoy, R.S. (1984), 'International Policy Implementation: A Theoretical Perspective', *Public Administration Review*, November\ December, pp.491-503.

Pag, S. and Wessels, W. (1988), 'Federal Republic of Germany', in Siedentopf, H. and Ziller, J. (eds) *Making European Policies work. The Implementation of Community Legislation in the Member States*, Sage, London.

Peterson, J. and Bomberg, E. (1999), *Decision-making in the European Union*, MacMillan, Basingstoke.

Pressman, J.L. and Wildavsky, A. (1974), *Implementation: how great expectations in Washington are dashed in Oakland*, Los Angeles, University of California.

Puchala, D. (1975), 'Domestic Politics and Regional Harmonisation in the European Communities', *World Politics*, vol.27, pp.496-520.

Rhodes, R.A.W (1988), *Beyond Westminster and Whitehall. The Sub-central Governments of Britain*, Unwin Hyman, London.

Rhodes, R.A.W. (1997), *Understanding Governance. Policy Networks, Governance, Reflexivity and Accountability*, Open University Press, Buckingham.

Rhodes, R.A.W. and Marsh, D. (eds) (1992), *Policy Networks in British Government*, Clarendon Press, Oxford.

Rhodes, R.A.W and Marsh, D. (1992a), *Implementing Thatcherite Policies. Audit of an Era*, Open University Press, Buckingham.

Richardson, J. (1996), 'Eroding EU Policies: Implementation Gaps, Cheating and Re-steering', in Richardson, J. (ed) *European Union. Power and Policy-Making*, Routledge, London.

Sabatier, P. (1986), 'Top-Down and Bottom-Up Approaches to Implementation Research: A Critical Analysis and Suggested Synthesis', *Journal of Public Policy*, vol.6, part 2, pp.21-48.

Sabatier, P. (1998), 'The advocacy coalition framework: revisions and relevance for Europe', *Journal of European Public Policy*, vol.5, No.1, pp.98-130.

Sabatier, P. and Mazmanian, D. (1980), 'The Implementation of Public Policy: A Framework of Analysis', *Policy Studies*, vol.8, pp.538-560.

Sabatier, P. and Mazmanian, D. (eds) (1981), *Effective Policy Implementation*, Lexington Books, Massachusetts.

Scharpf, F. (1994), *Community and Autonomy. Multi-level Policy-Making in the European Union*, European University Institute, Florence, Working Paper RSC No.94/1.

Schumann, W. (1991), 'EG-Forschung und Policy-Analyse. Zur Notwendigkeit, den ganzen Elefanten zu erfassen', *Politische Vierteljahresschrift*, 32.Jahrgang, Heft 2, pp.232-257.

Siedentopf, H. and Ziller, J. (eds) *Making European Policies work. The Implementation of Community Legislation in the Member States*, Sage, London 1988.

Weale, A. (1996), 'Environmental rules and rule-making in the European Union', *Journal of European Public Policy*, vol.3, No.4, December, pp.594-611.

3   Environmental Politics and Policy in the EU

Introduction

The following Chapter focuses on the first *layer* of the EU environmental policy process. In order to gain a comprehensive picture, the Chapter starts with an analysis of the EU policy-making process before investigating EU environmental policy objectives and their implementation problems. The Chapter describes EU policy-making as an enormously complex process which has been influenced by a multitude of actors with varying interests and which has culminated in a broad EU environmental policy containing some vague environmental policy compromises but also some substantial and ambitious environmental objectives. The EU has produced an impressive range of EU environmental policies, yet many policies have failed on the implementation ground. The Chapter addresses the apparent implementation deficit and investigates the imbalance between EU environmental policy ('over-') production on the one hand and national and *sub*national implementation shortfalls on the other. The *multi-layered implementation map* helps identify the key problems in implementing EU environmental policies:

- EU environmental policies often fail their *targets* because their legal instruments have limited direct binding force on implementors
- their *filtering* process is enormously complex and implementation links between the *layers* are weak
- the *layers* and their political-administrative systems are diverse, complex and feature formal and informal *determinants* which are often incompatible with EU environmental policies.

The Complexity of EU Environmental Policy-Making

EU environmental policies derive from a particularly complex policy-making process.[1] EU environmental policy-making involves not only EU

50

institutions,[2] but also national, *sub*national, and even international actors who seek to influence the form and content of EU environmental policies (see *feed-back* and *external factors arrows* in figure 3.1 below).[3] During the process, bargaining not only exists between government *layers* (vertically) but also within government *layers* (horizontally, cross-sectoral). In addition, institutions themselves have to come up with decisions which may have caused internal conflicts previously.[4]

In general, the Commission initiates an environmental policy as a response to informal pressures, i.e. demands from both environmentalists and representatives of economic level-playing-field considerations. Issued with information from the European Environment Agency (EEA) and lobbied by a vast variety of national and *sub*national institutions, interest groups and 'experts', the Commission prepares a policy draft which is then referred to the Council of Ministers and the European Parliament (EP) for consideration and decision-making. In addition, the Economic and Social Committee (EcoSoc) and the Committee of the Regions (CoR) are consulted whenever proposals affect their policy spheres. Depending on Member States' political and constitutional backgrounds, *sub*national representations formally participate in the EU environmental policy 'bargaining', too.[5] In the past EU environmental policy proposals were adopted through either the consultation and co-operation procedures (where the Council plays the key legislative role), or through the co-decision procedure (where the Council and the EP share legislative powers). The choice of procedures depended on the nature and importance of environmental issues. With the ratification of the Amsterdam Treaty in 1999, decision-making was simplified; a more streamlined co-decision procedure now applies to most EU environmental policies.

Throughout the process, politicians, administrators, experts, environmentalists and representatives of the farming and industrial sectors seek to influence the final policy version to suit their particular views and interests. Actors involved in the process pursue economic, environmental or political-strategic interests, and are influenced by events such as pollution incidents or economic crises. Considering the multitude of actors and their backgrounds as well as the variety and inter-connectedness of interests, EU environmental policy-making can either give way to a comparatively strong force in the bargaining process (e.g. a Member State or an EU institution),[6] or it results in a balance between conflicting sides and compromise solutions. In the latter case, the balancing act often leads to vague policy commitments which ultimately disappoint the majority of actors. The exception, perhaps, is the Commission which mediates between diverging interests and tends to consider any compromise a success. Put simply, environmental policy compromises are generally perceived as too stringent by business representatives, while the same policies are considered inadequate by environmental campaigners.

The search for environmental policy solutions is further complicated by the fact that the EU is equally committed towards common policy areas such as transport, trade and energy, which often compete with environmental objectives. Moreover, the EU established the principle of environmental policy 'integration' which implies, for instance, that a policy generating industrial production cannot be processed without taking into account aspects such as pollution, land use and energy efficiency. By the same token, environmental policy 'integration' also implies that environmental measures quantifying upper limits for pollutants cannot be set without taking account of the economic costs. Formulating an environmental policy proposal for consideration is therefore a delicate task for all actors involved.

Having negotiated an EU environmental policy with the EP, Council Ministers adopt a policy for a variety of reasons. They either fully support the environmental policy as it stands, they accept the policy in order to pursue other priorities in a policy 'package', or they accept the policy knowing that certain obligations can be avoided at a later stage. Individual Council Ministers may also be outvoted by their colleagues following a qualified majority voting procedure.[7] Policy decisions are therefore influenced by informal political-strategic calculations and formal procedures, which are relevant at the time of negotiation and may have a significant impact at a later implementation stage.

The adoption of the EIA Directive[8] and the preparations for its successors (Directive 97/11/EC amending the EIA Directive; the Strategic Environmental Assessment Directive) exemplify the continuing battle to find policy solutions. The EIA Directive was intended to harmonise and strengthen environmental standards in planning, a policy area which had previously been an exclusive domain of national and *sub*national actors. It took EU policy-makers ten years of intense discussion and considerable opposition from some Member States before the policy could be adopted. The UK Government saw the Directive as further 'red tape' preventing economic development, while the Danish Government was concerned about the transfer of their strict planning regulation powers to the European level. The EIA version, finally adopted in 1985, did not resemble the earlier policy draft which suggested more radical EIA measures. Instead, the Directive included a number of discretionary elements which provided the

Member States and their implementors with convenient planning policy loopholes.[9] Similarly, the 1997 amendment of the EIA Directive, forwarded by the Commission, met with considerable resistance, particularly from the German Minister for the Environment who rejected the policy proposal on the grounds that the new policy would have an 'inflationary' (and therefore counterproductive) impact on the original objective of the policy.[10] In the meantime, the EP expressed its disappointment that the new policy was not 'inflationary' enough - it did not comprise all the measures suggested by the EP. The SEA Directive has faced problems, too. In the latter case, *sub*national representatives stood at the forefront of opposition: in June 1997 the German Länder (through the Bundesrat) asked the Federal Government to reject the proposal. The Länder expressed concerns over accommodating the SEA policy into their existing legal-administrative systems and questioned the necessity to adopt a common SEA policy.[11] While the Commission and many EIA experts (i.e. planning officials and analysts) have insisted on further harmonisation in this policy area (which includes strategic planning), Länder officials continue with their campaign against the SEA Directive.[12]

It remains to be seen to what extent the German Länder will succeed with their campaign in a policy-making system, which is so complex in terms of actors, government *layers* and interests. The above examples illustrate, however, the tensions during policy-making and the difficulties in reaching policy solutions acceptable for all actors. Despite these difficulties, the EU has managed to establish a common ground on which a variety of environmental policies have been developed. The following Sections look at the evolution of the EU environmental policy in the Treaties and *Environmental Action Programmes* (*EAPs*) and highlight the multitude and magnitude of policy 'obligations' for national and *sub*national implementors. The Sections investigate to what extent the concerns and problems of national and particularly *sub*national implementors are taken into account in the Treaties and *EAPs*.

Figure 3.1: The EU Layer and Environmental Policy

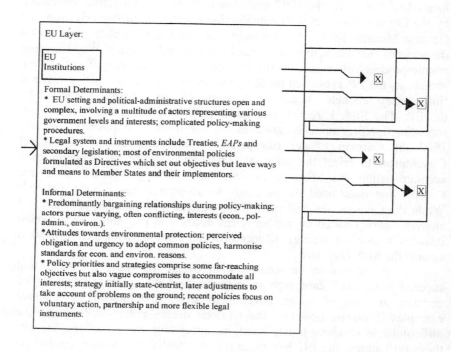

EU Layer:

EU Institutions

Formal Determinants:
* EU setting and political-administrative structures open and complex, involving a multitude of actors representing various government levels and interests; complicated policy-making procedures.
* Legal system and instruments include Treaties, *EAPs* and secondary legislation; most of environmental policies formulated as Directives which set out objectives but leave ways and means to Member States and their implementors.

Informal Determinants:
* Predominantly bargaining relationships during policy-making; actors pursue varying, often conflicting, interests (econ., pol-admin., environ.).
* Attitudes towards environmental protection: perceived obligation and urgency to adopt common policies, harmonise standards for econ. and environ. reasons.
* Policy priorities and strategies comprise some far-reaching objectives but also vague compromises to accommodate all interests; strategy initially state-centrist, later adjustments to take account of problems on the ground; recent policies focus on voluntary action, partnership and more flexible legal instruments.

The EU Environmental Policy in the Treaties

The manner in which a common environmental policy was established in the EU *layer* confirms the argument that a complex mix of *determinants* contributes to EU policy processes. Environmental considerations were not among the common policy priorities at the start of the European integration process, nor was the (then) EC the initiator of international co-operation in the environmental policy area. A significant international event preceded the EC's adoption of the environmental issue: the UN Conference on the Environment in Stockholm of June 1972 acknowledged the problems of pollution and their transboundary impacts. While the participants of the Stockholm Conference reinforced their sovereign right 'to exploit their own resources pursuant to their own environmental policies' as long as damage was not caused to other states, the conference, nevertheless, signalled a starting point in the international co-operation on environmental problems. EC Member States felt obliged to follow this high profile occasion (see *external factor arrow* in figure 3.1 above). In addition, Member States' governments noticed that domestic pressures for environmental policies and legislation would lead to diverging environmental standards which in turn would have an impact on trade within the EC. As a consequence, the EC soon adopted an *Environmental Action Programme* (*EAP*) which outlined a considerable number of environmental policy objectives and slowly adapted the treaty framework to environmental policy demands. *Sub*national views were not directly involved in the initial stages, neither did EU policy-makers pay full attention to possible implementation problems on the ground. EC environmental policies were considered at that time as foreign policy matters which formed part of the harmonisation process of national standards. In other words, EC environmental policies were developed and processed in a strictly intergovernmental and state-centrist manner. This shortfall would later be reflected in many disappointing EC/EU environmental policy outcomes.

It is striking how, in the years preceding the Single European Act, EC Member State governments justified the adoption of common environmental policies without changing the actual Treaty. The EC interpreted liberally a clause in the Rome Treaty Preamble which stated that the 'essential objective' of the EC was to improve living and working conditions for the Member States' citizens. As Haigh (1984) would put it, without amending the treaty text, the EC gave:

a gloss to the words of the Treaty and assume[d] that environmental policy was implicit.

Environmental legislation was processed under the economic provisions of Articles 100 and 235 RT,[13] which ensure a well-functioning Common/ Single Market. Applying Articles 100 and 235 RT, the objective was to coordinate and harmonise Member States' environmental standards and thereby avoid any distortions in the EC trade. This link between environmental measures and economic interests frequently resulted in rather curious legislative constructions. For instance, Freestone (1991) points out that the Directive on the conservation of species of wild birds was adopted under Common Market provisions.

The Single European Act

With the Single European Act (SEA), the (then) EC system was formally adjusted to allow for a legal basis for environmental policies under Articles 130R, 130S and 130T. Article 130R committed the EC and its Member States to the following objectives:

- to preserve, protect and improve the quality of the environment
- to contribute towards protecting the human health
- to ensure a prudent and rational utilisation of natural resources.

In this context, the SEA did not mention *sub*national actors and their vital role in the pursuance of these environmental objectives.
Environmental action was based on three principles:

- preventive action
- damage should be rectified at source
- the 'polluter pays principle'.

Further, the title stipulated that environmental protection requirements should be a component of the Community's other policies. In pursuing environmental policies, the EC should take into account:

- available scientific and technical data
- environmental conditions in the various regions of the Community
- costs and benefits of action\ non-action
- the economic and social development of the Community as a whole.

Paragraph 5 of Article 130R called upon the EC and its Member States to co-operate internationally with third countries in environmental matters. Article 130S identified the environment as a specific legislative area and specified that unanimity on environmental legislation was required in the Council of Ministers unless it was accepted that a qualified majority was sufficient for decision-making.

The SEA recognized diverse interests and conditions in the EC in two respects. Firstly, Article 130T allowed Member States to adopt more stringent protective measures as long as they were compatible with other objectives of the Treaty such as free and fair competition. Secondly, and more importantly, Paragraph 4 of Article 130R acknowledged the principle of subsidiarity and thereby distinguished between different government levels and their suitability for various environmental policy matters. However, the latter Paragraph fell short of specifying in detail the competencies for each government level (or *layer*). Moreover, it did not formally acknowledge the importance of the *sub*national level in the EC environmental policy process. In other words, whilst the SEA strengthened consistently the EC environmental remit, the important role of *sub*national regions in the overall process was not incorporated properly.

The Treaty on European Union

The Treaty on European Union (TEU), or Maastricht Treaty, elaborated on the SEA provisions but also altered the overall environmental policy picture. It integrated the environment more strongly into the EU formal framework: Article 2, which outlined the basic principles of the EU, now stressed the promotion of a:

> harmonious and balanced development of economic activities, sustainable and non-inflationary growth respecting the environment.

One TEU amendment was not directly related to environmental considerations but had implications for the EU environmental policy. Following increased pressures from several *sub*national governments in the Member States (in particular German Länder) and (for different reasons) the UK central government, the principle of subsidiarity was anchored more strongly into the overall Treaty framework. The new Article A stated that decisions should be taken as 'closely as possible to the citizen' and Article 3B called upon the EU and the Member States to take decisions at the most appropriate government levels.[14] In the context of environmental policies, subsidiarity spun off an ongoing debate on which government level is 'most appropriate' for decision-making in certain environmental policy (and related) areas. While subsidiarity has helped underline the importance of *sub*national participation in EU environmental politics, the provision did not resolve questions over competencies. Rather, it contributed towards greater divergencies of environmental policy commitment and performances. In fact, some Member State (and *sub*national) governments have pursued their stringent environmental measures, while others could justify their less stringent policies on the grounds of subsidiarity and diversity thus contributing to the overall implementation deficit in the EU environmental policy area. Detailed measures clarifying and regulating the principle of subsidiarity were missing from the TEU.

The TEU's amended Article 130R expanded on key environmental objectives. The EU underlined:

- its international commitment on environmental protection
- the principle to aim at a high level of environmental standards
- integrating environmental considerations into other EU policy areas.

Article 130S also provided rather complicated policy-making guidelines which specified the different legislative procedures depending on the form and content of proposals.[15] Article 130S(4) placed the financial burden as well as the implementation of environmental policies on the Member States unless measures were specified as common projects. Article 130S did not address *sub*national actors who actually implement the bulk of EU environmental policies. Legally, Member State governments' were to ensure that the policies are implemented and complied with by actors within the *sub*national regions. The supervision of *sub*national regions and their responsibilities in the EU environmental policy process were not explicitly mentioned in the TEU.

Article 130S(5) indicated a major change in the EU environmental policy: if EU environmental measures imposed disproportionate financial costs on one of the less prosperous EU regions (i.e. regions below 90% of the EU GDP average), Member States concerned could either derogate from this measure or apply for financial help under the Cohesion Fund. The latter provision allowed for more flexibility and acknowledged implementation problems faced by poorer Member States. On the other hand, the provision represented a potential loophole for Member States reluctant to cover the economic costs of EU environmental policies. As a final provision, Article 130T confirmed the right for individual Member States to adopt more stringent environmental measures as long as they were compatible with the Treaty. The Commission would have to be notified of these measures.

The Amsterdam Treaty

The Amsterdam Treaty (AT) of 1997 made minor adjustments to the EU environmental policy. The Treaty formally acknowledged sustainable development as one of the EU's key principles in Article 1. The Treaty also formally recognised environmental policy 'integration': new Article 3C stated that:

> [e]nvironmental protection requirements must be integrated into the definition and implementation of the Community policies and activities referred to in Article 3, in particular with a view to promoting sustainable development.[16]

Articles 130R, S, T were renumbered into Articles 174, 175 and 176 respectively and most EU environmental policies are now processed under the co-decision procedure. New Article 174 stated that:

> harmonisation measures answering environmental protection requirements shall include, where appropriate, a safeguard clause allowing Member States to take provisional measures, for non-economic reasons, subject to a Community inspection procedure.

Obviously, the latter provision was a response to the criticism of EU 'over-regulation' (or in German: Überreguliertheit)[17] and provides Member States and their implementors with more discretionary room to follow their EU environmental policy obligations.

On a wider policy scale, a Protocol on the principles of subsidiarity and simplification addressed general problems of the EU policy process. The new subsidiarity provision emphasised that only those policies were to be adopted at the EU level that were absolutely necessary. The protocol was intended to strengthen the legitimacy of EU policies and ultimately implementors' commitment towards EU policy objectives. Simplification of policy-making procedures and legislation would help EU citizens and implementors understand the EU policy process and thereby facilitate the implementation of EU policies. Apart from the protocol there were no other provisions in the AT which specifically dealt with implementation problems of EU policies in general and EU environmental policy problems in particular.[18]

The Treaty of Nice

With the forthcoming Eastern enlargement and unresolved questions regarding issues such as multi-speed integration, a common security policy and a new Charter on fundamental human rights, the environment was not high on the IGC agenda that led to the (provisional) adoption of the Treaty of Nice in December 2000. In fact, in terms of environmental policy change, the Nice Treaty has nothing substantial to offer. Member States have focused on a revised Article 175, which calls upon the EU Commission and the Member States to look into three areas:

- fiscal instruments
- measures affecting town and country planning, quantitative management of water resources and land use
- measures 'significantly affecting' the Member States' choice between energy sources and supply structures.

These policies, however, are to be adopted not under the co-decision procedure but under a separate provision that would allow the EP only consultative powers and require unanimity within the Council. The Treaty also includes a Declaration which stresses the EU's:

> (…) leading role in promoting environmental protection in the Union and in international efforts pursuing the same objective at the global level.

The Declaration also emphasises the need to use market-based incentives and instruments to promote sustainable development, a consideration included obviously with fiscal instruments (such as eco-taxes) in mind.

In sum, the EU has responded slowly to environmental problems, events and pressures (see *feed-back* and *external factors arrows*) and has developed an environmental policy, which now features some 'lowest-common-denominator' compromises but also some far-reaching principles and policy objectives. Over the years, the Treaties have adjusted and refined the EU environmental policy with sustainable development now playing a central role. At the same time, recent Treaties have introduced new measures which allow for more flexibility and discretion but also provide loopholes to avoid 'inconvenient' EU environmental standards. The most recent Treaty of Nice added the rather fashionable fiscal instrument of eco-taxes and incentives, presumably to explore further avenues other than simple (over-) regulation. In terms of diversity and variations in environmental quality and needs, the Maastricht and Amsterdam Treaties have briefly touched upon the Member States and regions' differences (evident in the principle of subsidiarity and the right to adopt more stringent policies) and have recognised some of the difficulties in meeting EU environmental policy obligations (the Cohesion Fund and the principle of simplification were intended to improve policy implementation). But these measures have not tackled the gap between policy-making and implementation by matching environmental principles and objectives with more systematic and legally binding implementation mechanisms tailored for both national and *sub*national *layers*. Neither have they changed behavioural patterns and committed national and *sub*national actors more strongly to EU environmental objectives.

If the Treaties themselves fail to fully address the problems of EU environmental policy implementation, perhaps EU *Environmental Action Programmes* (*EAPs*) compensate for this insufficiency. The following Section outlines the development of EC/EU environmental policy objectives as stated in the *EAPs* and assesses whether they address (and solve) implementation problems in the national and *sub*national *layers*.

EU Environmental Action Programmes

EU environmental policy priorities and strategies have been stipulated in *EAPs* from the early 1970s onwards. Following the events of the UN Stockholm Conference in 1972, the *first EAP* of 1973 established for the first time a common environmental policy basis and outlined the main objectives and principles which were re-confirmed in subsequent *EAPs* and, indeed, in the Treaties. The objectives were to:

- prevent, reduce and as far as possible to eliminate pollution and nuisances
- maintain a satisfactory ecological balance
- ensure sound management of natural resources
- improve the quality of life
- promote international co-operation.[19]

Environmental policies followed principles such as:

- pollution prevention
- preservation of natural resources
- polluter pays principle
- international commitments and responsibilities
- environmental education and awareness
- acknowledgment of the diversity of pollution as well as geographical and other differences
- possibility for individual Member States to adopt more stringent environmental policies.

Depending on policy objective and type, some of the *EAP's* objectives were to be carried out at the EC level, while many others required implementation within the 'domestic' boundaries of the Member States. Member State governments were required to supervise the policies' execution within their boundaries, while the Council further coordinated and harmonised national policies. The Commission was responsible for monitoring *EAP* compliance and for initiating further policy proposals.

The first *EAP* clearly focused on the intergovernmental harmonisation of national policies and neglected *sub*national variances. Moreover, the *EAP* was rudimentary in terms of structure and content with emphasis being placed on scientific research, the development of assessment criteria for future legislation and on specific issues such as the pollution of the Rhine.[20] Despite these shortcomings, the first *EAP* represented an important starting point in the EC/EU environmental policy process as it formulated for the first time principles and objectives which later contributed towards the adoption of common environmental policies.

The *second EAP* (1977)[21] built on its predecessor. One of its more significant features was its mentioning of the preventive policy approach. It also included specific policy issues such as environmental impact assessment (mentioned for the first time) and waste management. However, its emphasis on the urgent need for research, collection of data and their evaluation, indicated that the problems of pollution had not been tackled with the first *EAP* and that considerably more work was required in this field. In addition, the second *EAP* emphasised the need for more policy harmonisation: the Commission was requested to 'compare national laws' and 'align' laws wherever possible. This statement signalled for the first time that environmental policies varied significantly and required increased coordination. While the second *EAP* encouraged a more harmonised environmental policy framework, it addressed only variances at Member State level while intra-state or *sub*national variances were not mentioned at all.

The *third EAP* (1983)[22] differed from its predecessors in style and content. It focused on the integration of environmental interests into other policy areas and highlighted specific pollution problems in the Mediterranean region (as a response to the Southern enlargement of the 1980s). This time, Member States were noticeably careful to emphasise geographical and social differences as well as economic interests. In other words, the third *EAP* recognised the necessity to reconcile national and, indeed, regional socio-economic priorities with environmental considerations. To reassure themselves that environmental objectives did not suffer under economic considerations, Member States emphasised that environmental policies:

> must be carried out without regard to the short-term fluctuations in cyclical conditions.

More importantly, the third *EAP* was the first document to openly acknowledge the increasing gap between policy-making and implementation. It stated that projects from previous *EAPs* had not been accomplished and that there was still a:

> discrepancy between the scale of the projects and the means available for implementing them.

The Commission was instructed to progress the monitoring of implementation and increase its effectiveness in enforcing environmental legislation. In addition, the third *EAP* considered the development of clean technologies as a way to reconcile environmental interests and business opportunities. It was hoped that many of the disappointing EC environmental policies would benefit from this reconciliation.

The *fourth EAP* (1987)[23] reinstated a more rigorous policy approach. It confirmed long-established objectives and principles but also signalled a change in attitude. The problem of inadequate policy implementation was highlighted in the preceding Council Resolution: the Council:

> underlines the particular importance it attaches to the implementation of Community legislation and invites the Commission to review

systematically the application and the practical effects of existing Community policy and to provide regular reports on this to the Council and the European Parliament so that an assessment of the effectiveness of such a policy can be made and, inter alia, useful guidelines for future proposals determined.

Notwithstanding the apparent gap between environmental policy objectives and 'reality', the fourth *EAP* suggested even stricter environmental standards and legislation. The document argued that measures which closed discretionary loopholes for implementors, would benefit the environment and would also provide opportunities for the economy and employment. Stringent environmental standards would stimulate the development of 'clean technologies' and would make the EC as a whole more competitive at the global trading level. Apart from environmental standards and quantitative limits to pollution in EC legislation, the fourth *EAP* also considered economic instruments in the form of incentives and deterrents, and tolerated state aids for environmental projects 'under certain circumstances'. The Commission was also asked to develop a civil liability system, an objective which has not been fulfilled to date because Member States fear the financial costs for their industries.[24]

The *fifth EAP* (1993)[25] was clearly a response to the criticism which had been building up over the years concerning the implementation deficit in the EC/EU environmental policy area. The rigorous regulatory tone of the previous *EAP* was considered ineffective and was abandoned with the fifth *EAP*. The new *EAP* represented a desperate call for alternatives. Entitled *Towards Sustainability*, the fifth *EAP* followed the Brundtland Report of 1987 and the Dublin Declaration of 1990 which adopted sustainable development as a central theme.[26]

As a lesson from past implementation experiences, the fifth *EAP* highlighted revised environmental policy approaches. For instance, the EU should not rely on legal instruments which set quantitative levels, rules and standards. The variety of legal instruments should be widened to include policy measures such as market-based incentives and disincentives (e.g. voluntary eco-labelling and eco-auditing). Further, the Commission was instructed to establish dialogue groups which would give interested parties in all government *layers* an opportunity to take part in information exchange and policy-making (i.e. strengthening the *feed-back arrows*). The strategy was shifted from 'thou shalt not' to 'let's work together', a new campaign which responded to the criticism by many national and *sub*national implementors that 'Eurocrats' were 'out-of-touch'. Moreover, the fifth *EAP* praised public participation and encouraged Member States and EU institutions to consider citizens' complaints 'less a nuisance than a resource' for sustainable development. Another proposal for improvement, which was formally included in the fifth *EAP*, was the creation of a European Environment Agency (EEA), which today collects and processes information and makes data available to interested parties.

Since its adoption, the fifth *EAP* has been under intense scrutiny by the Commission. In November 1994, the Commission initiated an elaborate assessment process culminating in an *Interim Review of Implementation*. The *Interim Review* examined the Programme's implementation progress under six headings:

- 'integration' of environmental considerations into other policy areas
- 'broadening' the range of policy instruments
- shared responsibility and partnership (for instance through dialogue groups such as IMPEL)[27]
- changes in attitudes and patterns of consumption and production
- effective implementation and enforcement
- international responsibility.

These aspects were investigated within the areas of the manufacturing industry (in particular SMEs), energy, transport, agriculture, tourism, and international co-operation. The Commission's study concluded with 'cautious optimism', at the same time the Commission highlighted insufficiencies in meeting policy objectives as well as the necessity for further improvements. The Commission complained about the persisting perspective that environmental considerations were contrary to economic interests. Businesses continued to make use of environmental resources without fully covering the costs of pollution and environmental deterioration. The conciliation of environmental concerns and economic interests, in particular the 'internalisation' of environmental costs as part of businesses' cost-benefit analyses, remained one of the Commission's key objectives. Similar conclusions were presented by the Commission in 1999 in its *Global Assessment of the fifth EAP*, which highlighted again key problem areas, but particularly addressed the problem of persistent attitudes and production/ consumption patterns that were unsustainable.

In essence, the fifth *EAP* was intended to close the gap between policy-making and implementation and sought to apply new policy approaches such as 'smart' legal instruments (to use Ingram and Schneider's term) and dialogue at an early stage of the policy process. The new approaches were considered more compatible and acceptable for all interested parties in the national and *sub*national *layers* and signalled a strategic change from strict and regulatory policies to more pragmatic and flexible environmental policies.

In January 2001 the Commission published a proposal for a sixth *EAP* which yet again seeks to change attitudual patterns and close the gap between environmental policy intention and reality.[28] Entitled *Environment 2010: Our Future, Our Choice*, the sixth *EAP* is a ten-year plan which establishes a new set of seven priority areas for action:

- policy strategy (including implementation and transparency)
- climate change
- nature and bio-diversity
- environment and health (including the control of chemicals, pesticides, air and water pollution)
- sustainable use of natural resources and management of wastes
- international dimension (including measures supporting candidate states)
- public participation and access to information.

In the priority area of strategy, the sixth *EAP* reiterates the need for the Commission to continue launching infringement proceedings against non-compliant Member States in order to improve implementation records and make policies more effective. In addition, the Commission would also have to develop a new strategy of 'name, fame and shame'. The latter measure of praising 'good' while shaming 'bad' Member States, would be supported by improved public access to information, more transparency, and a new (and more effective) system of reporting, auditing and evaluation of environmental impacts of economic and other activities. For this purpose the Commission would explore further the possibility of environmental indicators that would apply throughout the EU.

Finally, the sixth *EAP* stresses the need to 'de-couple' negative environmental impacts from economic growth and establish an 'optimal' balance of environmental, social and economic objectives resulting in sustainable development. In order to accomplish this aim, the sixth *EAP* relies heavily on eco-efficient production methods, 'cleaner' (e-) economies, 'greener' lifestyles and informed consumer choices.

Overall, *EAPs* have corresponded more strongly with the practicalities of EU environmental policy objectives than the Treaties. They have set out EU environmental policy objectives, provided policy guidance for all actors involved and have been influential in so far as they paved the way for the adoption of environmental Directives and Regulations. On the other hand, *EAPs* have not carried the same legally binding weight as the Treaties. Despite a regular input of new policy ideas, strategies and encouragement to create a clean and healthy environment, *EAPs* have not generated the changes in behaviour and attitudes (i.e. informal *determinants*) envisaged by the authors of the *EAPs*; changes that would have contributed towards a more sustainable environment in the EU. Whether or not the forthcoming sixth *EAP* and its new strategy and prioritisation of environmental matters will make a difference in production and consumption patterns in the EU remains to be seen.

Table 3.1: Overview of Key EU Legal Instruments

Treaties (Primary Legislation):

Initially, EEC Treaty Articles 100 and 235 were applied for the adoption of EC environmental legislation.

Single European Act Articles 130 R, S, T set environmental objectives, principles and policy-making procedures; the Treaty allowed for Member States' stringent measures and acknowledged subsidiarity.

Treaty on European Union Article 2 included an environmental objective; the principle of subsidiarity was anchored more firmly into the Treaty framework; new Article 130R expanded on environmental objectives; 130S introduced more complicated policy-making procedures; the Treaty also mentioned responsibilities, introduced the Cohesion Fund, and confirmed the right to adopt more stringent measures.

Amsterdam Treaty formally acknowledged sustainable development as one of the key EU principles in Article 1; it stressed sectoral policy integration, and provided separate protocols on simplification and subsidiarity.

Treaty of Nice does not significantly change EU environmental policy, instead a revised Article 175 focuses mainly on fiscal instruments; a Declaration attached to the Treaty reinstates the EU's leading role in promoting environmental protection inside *and* outside the Union.

Environmental Action Programmes (EAPs):

First EAP (1973): starting point; outlined for the first time environmental objectives and principles; was rudimentary in some parts; based on intergovernmental harmonisation of national policies.

Second EAP (1977): followed format of first *EAP*; expressed the need for further research; included some specific target areas; emphasised comparison and alignment of national policies and the need for more harmonisation.

Third EAP (1983): included some specific issues such as Mediterranean problems; emphasised diversity; acknowledged the implementation gap and the need to monitor progress; expressed hope in clean technologies.

Fourth EAP (1987): signalled a more rigorous approach; stricter quantitative standards seen as opportunity to expand on new technologies; focused on economic instruments.

Fifth EAP (1993): dramatic response to the implementation gap; main theme sustainable development; new emphasis on partnership, participation and flexible legislative tools; progress on *Fifth EAP* reviewed extensively.

Sixth EAP (2001): signalling, yet again, a new strategy, focusing on a new set of priority areas, introducing the 'name, fame and shame' strategy and emphasising market-based solutions to environmental problems.

Types of Environmental Directives and Regulations (Secondary Legislation):

Procedural harmonization, e.g. EIA Directive (85/337/EEC); Directive amending EIA Directive (97/11/EC); IPC Directive (96/61/EC).

Setting qualitative/ quantitative standards, e.g. Directive (76/160/EEC) on the quality of bathing waters; Directive (80/778/EEC) on drinking water; Directive (85/203/EEC) on air quality.

Protection of species and areas of special interest, e.g. Habitats Directive (92/43/EEC); Wild Birds Directive (79/409/EEC).

Economic/ fiscal incentive policy, e.g. eco-label award scheme Regulation (880/92).

Other, e.g. Regulation setting up the European Environment Agency (1210/90).

Problems of EU Environmental Policy Implementation

Considering the complexity of government levels, actors and interests involved in the EU policy-making process, the EU has developed an impressive environmental policy. On the basis of the Treaties and the *EAPs*, the EU adopted more than 280 items of environmental legislation by the early 1990s of which about 200 Directives specified and formalised in more detail EU environmental policy objectives.[29] However, in contrast with the creation of EU environmental policies, their implementation has been rather disappointing (for supporting evidence see Table 3.2 below).

For a number of reasons, EU environmental policies have not been *filtered* through properly to the practical implementation *layers* with the result that implementors have often missed the original policy *targets*. Despite recent attempts to close the gap (attempts include dialogue groups and more flexible legal instruments) the EU is facing a serious implementation deficit in the environmental policy area. According to Krämer (1997):

> There is almost no other sector (...) where the gap between political statements and legislative commitments and obligations on the one hand and the reality [on the other hand] is as great as in the field of environment.

Why does the EU environmental policy area suffer from a gap between policy intention on paper and policy 'reality on the ground'? And why is there no other EU policy sector where the gap is as great as in the field of environment? The following analysis of key problems addresses the questions and sheds light onto the EU environmental policy practice.

(1) EU Environmental Policies fail to reach their Targets because their Legal Instruments are weak

Starting from the outset of policy implementation, the EU's legal instruments which outline EU environmental objectives (or policy *statements*, see *map*) imply problems for the policy process. EU policy-makers have at their disposal a wide range of EU tools[30] which help accommodate diverse policy matters, objectives and circumstances, but also cause confusion and provide loopholes for implementors. Directives, in particular, outline broader policy aims but also provide discretionary room for national and *sub*national implementors over the policies' ways and means. This discretionary room has often been used to avoid policy obligations. While this problem has been evident in other EU policy areas such as competition and social policy, the discretionary room for EU environmental Directives has been particularly large, partly because the Directives have had to accommodate several (long-term) objectives which affected several government levels and departments. In the case of Directive (76/160/EEC) on the quality of bathing waters, UK and German legislators used their discretionary room to interpret the notion of 'significant number of bathers' in such a way that only a small number of bathing waters were identified for monitoring. Following heavy criticism from the Commission, both the UK and the Germany eventually adjusted the number of their bathing waters from 27 to 470 and from 97 to 2000 respectively.[31]

Even within the category of Directives, the EU has developed a range of legal instruments with different obligations and control measures.[32] The variety of instruments allows for the political, economic and geographical diversity of the EU as well as the wide range of environmental policy matters. On the other hand, the complexity of EU instruments has had the effect that many implementors in the national and *sub*national *layers* have found it difficult to distinguish between instruments and interpreted their content incorrectly. Implementors have had difficulties with the details of EU environmental laws which often contained several objectives and affected several policy areas. Moreover, the study of EU environmental laws has been time consuming for implementors who have rather dealt with familiar matters first before attending to complicated, and sometimes 'inconvenient', EU environmental policy documents.[33] Aware of implementors' difficulties with EU legal instruments, the Commission has suggested that documents and legislative processes should be simplified and streamlined as much as possible.[34] However, the 1990s have seen more complex environmental policy demands such as the integration of environmental considerations into other policy areas and the adoption of procedural, cross-sectoral policies such as the IPC Directive. These policies have been difficult to accommodate in the light of the Commission's simplification strategy. The new approach of simplification has therefore been limited to certain environmental objectives.

Another major problem of EU legal instruments concerns the actual language used in formulating EU environmental policies and their subsequent interpretation. Following an elaborate bargaining process, Directives in particular often feature legal formulations which are deliberately vague and open to interpretation to suit all interests involved. General terms such as 'best available technology not entailing excessive costs', 'high environmental standards' or 'no nuisance' are not sufficiently defined to provide for a truly common environmental policy basis. The term 'best available technology not entailing excessive costs' (known as the

BATNEEC principle), for instance, raises questions such as - which technology is 'best'; when are costs 'excessive'; and who decides whether costs are excessive or not? Apart from the problems of definition and interpretation of certain terms, Directives provide large discretionary powers over the ways and means of policy implementation. This lack of clarity has encouraged Member States and their implementors to interpret EU legislation to their own liking with the result that many EU environmental policy *targets* have not been met on the ground.[35]

(2) The Filtering Process of EU Environmental Policies is enormously complex and the Links between Layers are weak.

Focusing on the actual implementation process, EU environmental policies require formal transposition by the Member States and their national and (depending on Member State) *sub*national legislators. In the case of Directives, national and *sub*national legislators are also required to specify policy requirements where the EU legal text provides flexibility and discretion. National and *sub*national administrators should then proceed with the implementation by establishing guidelines and by applying the policy on the ground.[36] Administrators are supported by various quasi-governmental bodies and environmental non-governmental organisations (NGOs) which contribute information and know-how. Citizens as well as representatives from industry, the farming community and environmental NGOs are expected to comply with the policy. Finally, courts within the Member States have the task of interpreting EU legislation and, if necessary, enforce full compliance with legal obligations.

Figure 3.2 below describes the *filtering* process of EU environmental policies as complex in many respects. Firstly, the vast majority of EU environmental policies (e.g. policies regulating water and waste management, pollution control policies, the EIA policy discussed in Chapter 6) require implementation at the national and *sub*national government levels. In other words, policies have to be *filtered* through essentially two *layers* before they reach their implementation *target*. Considering the long distance between policy *statement* and *target*, EU environmental policies face more potential implementation obstacles than, say, national policies.

Secondly, the *filtering* process of EU environmental policies involves a multitude of actors who are influenced by a variety of informal *determinants* (i.e. attitudes towards environmental protection and the EU, policy-makers' priorities and strategies). Following Offe's concept of subsystems, EU environmental policies tend to suffer from tensions between representatives of economic interests (e.g. business and farming communities) and environmental interests (e.g. environmental NGOs) while political-administrative actors (e.g. Commission officials and administrators on the ground) try to mediate between the two groups. The 'push-pull' effect of conflicting interests hinders the implementation process. In the case of the EIA Directive, planning officials in the national and *sub*national *layers* have tried to consolidate the developers' economic interests with environmental concerns of NGOs and citizens affected by project applications. During the balancing process, planning officials have often opted for the easiest solution and have given way to economic interests. As a result, the key objective of the EIA Directive (i.e. environmentally sound 'minimum-regret-planning') has often been ignored in practice.[37]

Thirdly, the *filtering* of particularly ambitious EU environmental policies often requires major adjustments of existing formal and informal conditions in the national and *sub*national *layers*. Formal *determinants* such as legal systems and political-administrative structures as well as informal *determinants* such as policy-makers' priorities and attitudes towards environmental protection in the national and *sub*national *layers* are not always compatible with EU environmental policies. In practice, favourable preconditions for successful implementation such as the commitment and flexibility of national and *sub*national actors have often been missing during the *filtering* process. Facing the inflexibility of national and *sub*national political-administrative structures and practices, many EU environmental policies have simply failed to reach their *target*.

Finally, the *filtering* process has been hampered by the way competencies have been allocated and shared between EU, national and *sub*national *layers* and their actors. According to the treaties, the Commission is required to 'guard' the implementation of policy obligations as stated in the treaties and EU legislation.[38] The Commission is supposed to 'remind' Member States of policy commitments and, if necessary, threaten Member States with fines and court action if they fail to fulfil their tasks or violate against legislation. While Member States are required to follow their EU environmental policy commitments, national and *sub*national implementors' responsibilities are not specifically addressed in the treaties. EU policy-makers have been careful not to dictate the allocation of competencies, not only because they are so diverse in terms of state systems and government structures but also because they have feared that the allocation of competencies would start a major debate on national-

*sub*national relations and power-sharing.

In practice, the Commission's guardian activities have almost always come to a halt at the first stage of implementation (i.e. formal transposition of EU legislation). Member States and (depending on constitutional settings and the policy in question) their *sub*national regions have often ignored deadlines or failed to notify the Commission of any changes in national (and *sub*national) legislation. In many cases, Member States and *sub*national regions have had difficulties in interpreting Directive objectives or they assumed that their own legal instruments and administrative structures and resources were adequate.[39]

If Member States fail to transpose and implement Directive objectives into the national context or fail to put pressure on their *sub*national regions to follow their EU obligations, the Commission can initiate infringement proceedings under Article 169 (new Article 226).[40] The infringement procedure is divided into three stages. First, the Commission informs the Member State of a suspected infringement which has come to its notice ('letter of formal notice') and requests the Member State to submit its observations. If the Member State's response is not satisfactory, the Commission issues a 'reasoned opinion' stating why infringement is suspected. If the Member State still does not show any reaction to the Commission's concern, the matter is taken to the European Court of Justice for judicial ruling.

In practice, the first stage has often clarified misunderstandings and implementation problems.[41] Both the Commission and the Member State governments have sought to resolve problems at an early stage because they did not wish to rock the boat. However, many cases have not been taken further by the Commission out of reluctance to 'over-use' the threat of legal prosecution. In addition, the Commission has been very careful not to appear as a dominant power 'from Brussels' for Member States and their implementors. Yet, despite this diplomatic approach, the Commission has had to initiate several legal proceedings against Member States in 1997. According to Smith (1997), thirteen out of fifteen Member States were challenged over their failure to comply with EU environmental legislation on water quality, waste management and nature protection (in particular the Habitats Directive).

If Member State governments refuse to comply with EU environmental policy obligations or are unable to commit their *sub*national regions to EU policy compliance, the cases are referred to the ECJ which is likely to impose fines under TEU Article 171 (new Article 228). It is debatable whether the Commission's legal challenge together with the fines imposed by the ECJ have a significant impact on the Member States' compliance with EU environmental legislation. The moral pressure caused by the publication of fines certainly has had some effect on the Member States and their implementors.[42] On the other hand, many Member States have been prepared to pay the fines since relatively small amounts of money are considered worthwhile in the light of costly environmental standards and economic difficulties.[43]

In order to fulfil its guardian task and strengthen the *filtering* links from one *layer* to another, the Treaties should provide the Commission with more effective powers enabling it to control and, if necessary, enforce compliance within the Member States and their *sub*national regions. Equally, Member State governments should have adequate monitoring mechanisms in order to check EU policy compliance within the *sub*national regions. However, neither the Member State governments[44] nor the Commission have shown effective controlling powers. The Commission in particular, does not possess adequate staff and financial resources to monitor the implementation of every EU environmental policy in every part of the EU. It does not possess 'eco-inspectorate' powers similar to the fisheries inspectorate functions under the Common Fisheries Policy. Instead, it has to rely upon complaints from environmental NGOs and individual citizens who inform the Commission of any alleged infringement cases.[45] Complaints, however, are insufficient in monitoring overall compliance. For instance, UK citizens have tended to complain more than citizens do from Denmark and within the UK there have been more complaints from England than from Scotland.[46] This imbalance contributes to an incoherent picture of the whole policy area.

Member States and their *sub*national regions have not been assessed on their performance in a uniform and transparent manner and therefore cannot be properly compared with each other. Further, implementors' discipline and commitment towards EU environmental obligations has been relatively low. To a certain extent, EU policy-makers have acknowledged the problem by creating another EU institution, the EEA, which has developed a system of uniform and comparable information gathering and dissemination. The system has helped to identify cases of non-implementation and has thereby 'embarrassed' those national and *sub*national actors who did not comply with EU environmental policies. However, while the EEA has processed environmental information, its tasks have not included an active 'eco-inspectorate' function. When the EEA was established, the Member States deliberately restricted its tasks to the gathering and distribution of information. Council

Ministers were not prepared to:

> have their performances vetted by another tier of 'Brussels bureaucrats'.[47]

Without a control body, however, it is not only difficult to gain a coherent and accurate overview of the success or failure of environmental policies, it is also difficult to fully enforce implementation discipline within the Member States and their regions.

Since the early 1990s, the Commission has sought to consolidate policy-making and implementation by initiating new, alternative policy strategies which more strongly involve actors from all government levels and interest groups. In particular, the Commission has focused on 'working' or 'dialogue' groups which include a wide spectrum of actors and their interests. The Commission now receives information on implementation performances from IMPEL, a forum for information exchange and dialogue which assesses the practicalities of EU environmental policies. Apart from dialogue and partnership initiatives, the Commission has responded to discrepancies between stated policy objective and 'reality' by setting policy expectations at a lower, more pragmatic level. Many new environmental policy proposals focus on voluntary, market-based solutions to environmental problems. As a response to national and *sub*national implementors' criticism of EU 'Überreguliertheit', the Commission has also considerably reduced the number of environmental policy proposals to allow government levels below to take their own environmental policy decisions (under the principle of subsidiarity). In addition, an internal Commission communication has suggested a systematic 'implementation check' for Commission. This mechanism assesses the EU environmental policies' potential costs and benefits as well as feasibility and compatibility with national and *sub*national legal-administrative systems. However, some Commission officials have already expressed doubts over the necessity and usefulness of such a 'check'.[48]

While the Commission hopes that these and other policy initiatives will help establish more acceptable EU environmental policies, it remains to be seen whether these efforts can really strengthen implementation links and make the *filtering* process more permeable. The Member States and their *sub*national regions have repeatedly demonstrated reluctance in accepting more binding monitoring and enforcement mechanisms. In the light of the weak links between the *layers*, the gap between policy *statement* and *target* is very likely to remain.

(3) The Layers involved in the Filtering Process are diverse, complex and their formal and informal Determinants are often incompatible with EU Environmental Policies.

EU environmental policies require implementation in national and *sub*national *layers* which are complex and diverse. In terms of formal *determinants*, constitutional settings, political-administrative structures and resources as well as legal systems (all of which are shaped by policy styles, practices and priorities) have varied across and, more importantly, within Member States.[49] Similarly, informal *determinants* such as relationships between actors, attitudes towards environmental protection and the EU, and policy-makers' strategies and practices (all of which are influenced by formal conditions) have varied, too. This diversity of *determinants* is understandable and legitimate. Nevertheless, the question arises whether some of the national and *sub*national *layers* can cope with EU environmental policies. Given the wide array of formal and informal *determinant* combinations, the ways in which EU environmental policies have been implemented (or not implemented) within the Member States and their *sub*national regions resemble a colourful patchwork. Many *layers* have featured an unfavourable combination or 'mix' of *determinants* which have hindered the successful implementation of EU environmental policies.[50] National and *sub*national implementors have also tended to amend EU environmental policies as much as possible to suit their particular 'mix' of formal structures and informal priorities.[51] As a result, EU environmental policies have often assumed a shape which has not resembled the policy intention at the outset of the process.

Focusing on formal *determinants*, EU environmental policies have often been hindered due to a lack of resources or inadequate administrative structures within the national and *sub*national *layers*.[52] Budgetary constraints in particular, have put a damper on EU environmental policy obligations and in many cases the EU Cohesion Fund did not compensate for the problems of financing the implementation of policies which demand, for instance, expensive technological standards. In addition, some Member States and their *sub*national regions have featured political-administrative structures which were less suited than others to cope with the requirements of EU environmental policies. For instance, they have not been equipped with the institutions and technological know-how necessary to measure quantitative environmental standards.[53] The details of EU environmental legislation (e.g. quantitative thresholds and qualitative standards) have also clashed with national and *sub*national legal systems and instruments. There are, therefore, a number of potential formal obstacles (highlighted in figure 3.2 below) which can hinder the successful implementation of EU environmental policies.

In terms of informal *determinants*, particularly attitudes towards the EU as well as policy-makers' economic priorities have played a significant role in the implementation of EU environmental policies. Firstly, political-administrative actors in the national and *sub*national *layers* have not received EU environmental policies unprejudiced. Often, disputes over conflicting interests which had been conducted 'out in the open' during policy negotiation have not been resolved with the adoption of a policy. Reacting to preceding conflicts, actors have not felt obliged to show much enthusiasm towards a controversial environmental policy.[54] In other cases, national and *sub*national actors have reacted towards EU environmental policies with either disappointment over 'lowest-common-denominator' compromises,[55] or they have been preoccupied with other policy priorities and have not taken much notice of EU environmental policies. Often, national and *sub*national actors have not been particularly biased against EU environmental policies but have chosen the easiest option which caused the least friction: they have avoided 'inconvenient' policy obligations for as long as possible.

In terms of relationships with, and attitudes towards the EU, many national and *sub*national actors have questioned the legitimacy with which EU environmental policies have been adopted. Particularly *sub*national implementors have often considered EU policies as 'imported legislation'[56] and have clashed with EU policy-makers over the question whether a certain environmental policy required harmonisation at the EU level. *Sub*national regions with a strong territorial identity such as Bavaria have perceived 'policies from Brussels' as a challenge against their own political competencies. Reassurances on the principle of subsidiarity in the Treaties and other EU documents have not eliminated the scepticism over 'unnecessary policies from Brussels' by many implementors on the ground.[57]

Not far from the legitimacy question is the economic imperative which influences EU environmental policies. Economic motivations which contributed towards the adoption of EU environmental policies have tended to differ from the economic (self-) interests of national and *sub*national actors on the ground. More specifically, economic concerns over possible imbalances in the 'level-playing-field' caused by diverging environmental requirements which compelled Member States' governments to harmonise their national environmental policies, have tended to evaporate at a later stage in the light of immediate interests in protecting and generating national and *sub*national economies. Increased trade has been a desirable but abstract goal. However, actors on the practical ground have found it difficult to believe that policies, which effectively restrict their economic activities, not only protect the environment but also benefit their economies in the long term. National and *sub*national implementors and business communities have therefore tended to secure economic priorities first before attending to the 'inconvenient' harmonisation of costly environmental measures. *Sub*national administrators in particular have shown a protectionist attitude towards their local economies and have been less sympathetic towards the goal of EU-wide environmental standards harmonisation (see Chapters 5 and 6). This discrepancy between the 'level-playing-field' objective in one *layer* and economic self-interests in the other *layers* helps explain why many EU environmental policy outcomes have been disappointing and why a wide gap is still evident between EU environmental policy *statement* and *target*. It underlines the importance to examine the motivations of those who are actually charged with the formal and practical implementation of EU environmental policies.

Figure 3.2:   General Implementation Obstacles for EU Environmental Policies

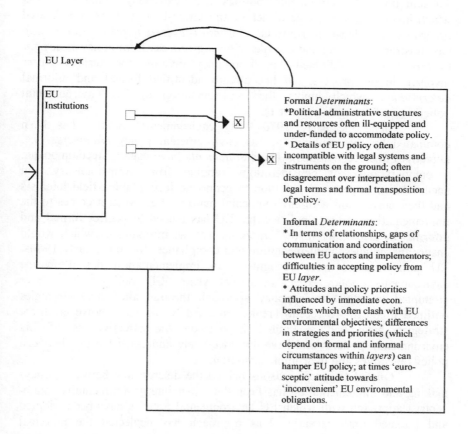

EU Layer

EU Institutions

Formal *Determinants*:
*Political-administrative structures and resources often ill-equipped and under-funded to accommodate policy.
* Details of EU policy often incompatible with legal systems and instruments on the ground; often disagreement over interpretation of legal terms and formal transposition of policy.

Informal *Determinants*:
* In terms of relationships, gaps of communication and coordination between EU actors and implementors; difficulties in accepting policy from EU *layer*.
* Attitudes and policy priorities influenced by immediate econ. benefits which often clash with EU environmental objectives; differences in strategies and priorities (which depend on formal and informal circumstances within *layers*) can hamper EU policy; at times 'euro-sceptic' attitude towards 'inconvenient' EU environmental obligations.

Conclusion: Between EU Environmental Policy Ambition and Reality

To sum up, EU environmental politics is an extremely complex process which involves a vast range of actors and interests as well as complicated procedures and legal instruments. The *multi-layered implementation map* has illustrated the long and complex *filtering* process from policy *statement* to *target*, has highlighted the discrepancy between the various *layers* involved in the process and has already identified formal and informal *determinants* which hinder the implementation of EU environmental policies (confirming Argument 1).

Since the early 1970s, EU environmental politics has been dominated by the 'making' of EU environmental policy *statements* (i.e. Directives and Regulations). These *statements* have usually accommodated a complex mix of considerations ranging from transboundary and accumulative impacts of pollution to economic level-playing-field interests and their correlation with environmental matters. Yet, when it comes to the implementation of these policies, the EU has tended to neglect formal and informal *determinants* in the *layers* as well as mechanisms which would ensure the policies' implementation and compliance. Since the early 1980s, EU policy-makers have recognised an implementation deficit in the environmental policy area. In recent years, EU policy-makers have attempted to adjust their policy approach through alternative strategies outlined in the more recent Treaties and *EAPs*. To date, however, these strategies have had a limited impact on the effectiveness of EU environmental policies and have not adequately addressed the gap between policy objectives and their implementation.

One of the main reasons behind the discrepancy between policy-making and implementation has been the rather intergovernmentalist, state-centrist approach with which EU environmental policies have been adopted and pursued until recently. This approach has neglected the practical problems of EU environmental policy implementation on the ground as well as the multi-facetedness of Member States and their intra-state variances. Most EU environmental policies require implementation at all government levels and particularly at the regional and local levels. Yet the complexity and influence of these levels on EU environmental policies have been to date under-estimated. *Sub*national regions are in many ways distinct from their 'mother' states and feature formal and informal *determinants* which can differ significantly from *determinants* which shape politics and policy processes in the Member States at large. During the *filtering* process, EU environmental policies can clash with incompatible *determinants* within the *sub*national regions; *determinants* which would not be detected with a state-centrist ('domestic') research method. Therefore, in order to gain a more accurate picture of the EU environmental policy reality (and close the implementation gap), it is necessary to investigate implementation practices more thoroughly and go beyond the national (or 'domestic') level to include specifically *sub*national conditions and obstacles during policy implementation.

Chapters 4 and 5 refine the study of EU environmental policy implementation by investigating the national and *sub*national *layers* separately. The Chapters assess to what extent *sub*national regions differ from their 'mother' states and to what extent they shape the EU environmental policies during implementation (addressing Argument 2). Chapter 6 will then examine in detail the *filtering* process through the *layers* by focusing on one specific policy example: the EIA Directive. The following Chapters will confirm that *sub*national regions and their actors do play a significant role in EU environmental politics and that they therefore deserve more attention by EU practitioners and researchers.

Table 3.2: Evidence on EU Environmental Policy Implementation

In its 1994 Annual Report to the European Parliament, the European Commission addressed the problem of insufficient implementation of EU environmental legislation. From the information available, the Commission listed the most common insufficiencies and highlighted implementation problems of individual Member States. According to the Commission, the most common problems were: delays in transposing Directives, incorrect transposal and incorrect application. In many infringement cases, the Commission had to rely on information from complainants. The Directives on 'freedom of access to information on the environment' (90/313/EEC) and 'environmental impact assessment' (85/337/EEC) deserved separate paragraphs in the Report since their insufficient implementation was particularly striking.

The *Fifth Environmental Action Programme* (*EAP*) outlined 'some disquieting trends' which, 'if not satisfactorily contained, could have significant negative consequences for the quality of the environment as a whole'. Trends included a 20% increase in EC carbon emissions by 2010 (reference year 1987), a 25% increase in car ownership and a 17% increase in mileage by 2000 (reference year 1990), a 63% increase in fertiliser use between 1970 and 1988, a 13% increase in municipal waste over the last 5 years, and a 60% increase in Mediterranean tourism projected by 2000 (reference year 1990).

The Commission's *Interim Review of Implementation of the Fifth EAP* concluded with 'cautious optimism' on the progress of environmental measures in the EU. However, the *Review* also emphasised areas of limited success and problems in ensuring a sustainable environment in Europe. Problems included insufficient awareness of pollution and environmental deterioration, lack of willingness to adapt to environmental demands, and a general attitude that environmental matters only concern those who work in the environmental sector.

In its 1996 *Summary of the Progress Report on the Fifth EAP*, the Commission stated the following: 'Member States ultimately determine the effectiveness of Union measures. There are delays and failures in properly transposing directives into national law, and in some cases failure to enforce compliance with the transposed law. Late transposition of legislation remains an endemic problem in a majority of Member States.'

The Institute for European Environmental Policy (IEEP) examined 'the state of reporting' by the Commission. According to IEEP, the Commission failed to report on the progress of EU environmental policy implementation for several reasons: in many cases legislation had been poorly drafted and caused confusion; Member States failed to comply with their reporting obligations; the Commission conducted cumbersome internal procedures; the Commission lacked resources and, in some cases, lacked the will to follow up a policy; and finally other EU institutions failed to put pressure on the Commission to fulfil its obligations. IEEP included a comprehensive list of environmental Directives requiring reports. According to the list, the Commission had not submitted a large number of reports at the time of publication.

Notes

[1] For a more detailed overview of key EU actors, procedures and legislation see Nugent (1999), Dinan (1999), Peterson and Bomberg (1999).

[2] The main policy-making institutions of the EU are - the Commission, the Council of Ministers, the European Parliament (EP), the Committee of the Regions (CoR) and the Economic and Social Committee (EcoSoc).

[3] Liberatore (1991) highlights the influence the international level (and more specifically the UN) has had on EU environmental policy.

[4] For instance, differences in views often occur between Commission DG XI (environment) and other DGs such as DG VII (transport). The same rule applies to single departments: even within Commission DG XI opinions and priorities can clash.

[5] Keating and Hooghe (1996) provide detailed information on *sub*national participation in EU policy-making and distinguish between various forms of regional influence.

[6] For instance, Germany was the driving force behind the adoption of the LCP Directive (which regulates emissions from large combustion plants). For a detailed account of German and UK attitudes towards air pollution and the LCP Directive see Böhmer-Christiansen and Skea (1991).

[7] Votes are allocated as follows: Germany, UK, France and Italy 10 each; Spain 8; the Netherlands, Greece, Belgium, Portugal 5 each; Sweden and Austria 4 each; Denmark,

Finland and Ireland 3; Luxembourg 2. Total number of votes 87, 62 votes required for qualified majority. In preparation for Eastern enlargement, votes and qualified majorities/ minorities will be re-adjusted once the Treaty of Nice is adopted and ratified.

[8] Council Directive on the assessment of the effects of certain public and private projects on the environment (85/337/EEC). The EIA Directive is discussed in detail in Chapter 6.

[9] For a detailed account of the Directive and its 'evolution' see Wood (1995).

[10] Former Federal Minister for the Environment, Angela Merkel (CDU), explained her decision to oppose the amendment in a letter of 20. December 1995 sent to the Länder Ministries for the Environment.

[11] See 'Bundesrat lehnt UVP für Pläne und Programme ab' UVP-Gesellschaft (http://www.laum.uni-hannover.de/uvp/uvp-netz/).

[12] At the time of writing the SEA Directive proposal was still with the Council of Ministers and the European Parliament for consideration. The Directive's adoption is expected to be completed by the end of 2001.

[13] Article 100: The Council shall, acting unanimously on a proposal from the Commission and after consulting the European Parliament and the Economic and Social Committee, issue directives for the approximation of such laws, regulations or administrative provisions of the Member States as directly affect the establishment or functioning of the common market. Article 235: If action by the Community should prove necessary to attain, in the course of the operation of the common market, one of the objectives of the Community and this Treaty has not provided the necessary powers, the Council shall, acting unanimously on a proposal from the Commission and after consulting the European Parliament, take the appropriate measures. The Treaty of Amsterdam renumbered Articles 100 and 235 as Articles 94 and 307 respectively.

[14] Art.3B (new Article 5): The Community shall act within the limits of the powers conferred upon it by this Treaty and of the objectives assigned to it therein. In areas which do not fall within its exclusive competence, the Community shall take action, in accordance with the principle of subsidiarity, only if and in so far as the objectives of the proposed action cannot be sufficiently achieved by the Member States and can therefore, by reason of the scale or effects of the proposed action, be better achieved by the Community. Any action by the Community shall not go beyond what is necessary to achieve the objectives of this Treaty.

[15] As a general rule, the more sensitive issues were (in particular those that may have had considerable fiscal and economic implications), the more difficult it was for the EP to influence the policy and the more difficult it was for the Council to reach a decision (in these cases unanimity was often required).

[16] The objective of environmental integration was previously mentioned in Art.130R(2). Since the new Art.3C now covers 'integration', the Art.130R(2) 'integration' clause has been deleted from the Amsterdam Treaty.

[17] For a discussion on the problem of 'Überreguliertheit' see Demmke (1996).

[18] The other Treaty amendment concerned EU decision-making: Art.130S specified that decisions on Art.130R matters should be taken under the co-decision procedure as outlined in Art.189B instead of the Art.189C co-operation procedure. With the Amsterdam Treaty, Articles were re-numbered as follows: ex-Art.130R - Art.174; ex-Art.130S - Art.175; ex-Art.130T - Art.176.

[19] The first *EAP* was presented by the European Commission in 1973 and received strong support from the Council and the Member States in a 'Declaration' which accompanied the document. See *O.J.* (1973).

[20] The pollution of the Rhine became evident at the end of the 1950s and was high on the political and media agenda throughout the 1960s.

[21] See *O.J.* (1977).

[22] See *O.J.* (1983).

[23] See *O.J.* (1987).

[24] The Commission considered this idea in its 1995 *Annual Programme* but is only now preparing a policy proposal on 'environmental liability'. Although the idea of liability has been circulating for quite some time, many Member States and business representatives are reluctant to accept a system which would imply substantial and unpredictable costs for their economies.

[25] See *O.J.* (1993).

[26] The *fifth EAP* cites the Brundtland definition of sustainable development. Sustainable development refers to the 'development which meets the needs of the present without compromising the ability of the future generations to meet their own needs'. The document did not, however, clarify questions such as - who determines or predicts the needs of future generations; who safeguards their interests; and what exactly are their needs in the future?

[27] IMPEL: EU Network for the Implementation and Enforcement of Environmental Law, formerly 'Chester Network'. For a detailed discussion of IMPEL see Werner (1996).

[28] The proposal has been forwarded to the Council of Ministers and the European Parliament for consideration and currently (i.e. at the time of writing) awaits adoption.

[29] The figures are taken from Young's (1993) analysis of EC environmental policy.

[30] A Regulation is directly binding for all 'in its entirety'; it should, ideally, be clear in its objectives and there should be no room for interpretations or legal loopholes. A Directive is binding as far as the final aim is concerned; ways and means are left to the Member States and their implementors. A Decision is binding in its entirety upon those to whom it is addressed; it is used to remind a Member State/ institution of certain obligations. Recommendations and opinions have no direct legal force; they are merely a political statement and call upon the addressee to follow EU obligations.

[31] Information taken from Krämer (1997).

[32] Rehbinder and Stewart distinguish between three Directive categories: the 'typical' Directive contains an obligatory goal but leaves the means to the Member States, the 'regulation-type' Directive contains detailed substantive obligations (standards, provisions for implementation etc.), and the 'framework' Directive paves the way towards a 'regulation-type' Directive. Cited in Freestone (1991).

[33] In the words of an Environmental Consultant, EU environmental policy documents tend to end up at the bottom of the 'in' tray. Interview, 15. February 1995, Dalkeith.

[34] This move towards simplification in the environmental policy area was highlighted in a Commission DG XI interview, Brussels, 7. March 1997; see also Commission (1996b) and McHugh (1996).

[35] Bakkenist (1994) describes in detail how UK actors interpret liberally Directive (90/313/EEC) on access of environmental information.

[36] Art.5 (new Art.10): Member States shall take all appropriate measures, whether general or particular, to ensure fulfilment of the obligations arising out of this Treaty or resulting from action taken by the Institutions of the Community. They shall facilitate achievement of the Community's tasks. They shall abstain from any measure which could jeopardise the attainment of the objectives of this Treaty. Art.130R (new Art.174) reinforces the Member States' responsibility of implementation.

[37] See Chapter 6 for a detailed analysis of the EIA Directive and its implementation in Scotland and Bavaria.

[38] The Commission has to 'ensure that the provisions of this Treaty and the measures taken by the institutions pursuant thereto are applied' (Art.155, new Art.211).

[39] For instance, Bavarian legislators argued that the formal transposition of the EIA Directive was not necessary because existing legal-administrative provisions in Bavaria were adequate. See Chapter 6 for a detailed account of the Directive's implementation.

[40] Art.169 (new Art.226): If the Commission considers that a Member State has failed to fulfil an obligation under this Treaty, it shall deliver a reasoned opinion on the matter after giving the State concerned the opportunity to submit its observations. If the State concerned does not comply with the opinion within the period laid down by the Commission, the latter may bring the matter before the Court of Justice.

[41] In general, most disputes are resolved during the first stage following the 'letter of formal notice' (1209 cases). The number of 'reasoned opinion' cases for the same period is considerably smaller (342 cases), while the number of references to the ECJ is comparatively insignificant (44 cases). The figures are taken from Butt Philip (1994) and refer to Community Law in general for 1993.

[42] In 1997, Germany faced heavy fines imposed by the ECJ for insufficient implementation of EU environmental legislation in three instances. Observers in Brussels saw the 'punishment' as an embarrassment and effective deterrent for Germany. For more details see *Europe* (1997).

[43] Butt Philip (1994) argues that current penalties are not threatening enough for polluters.

[44] Reporting by *sub*national actors on the progress of EU environmental policy implementation is predominantly on an informal and voluntary basis despite specific reporting requirements outlined in EU Directives. For further information on the difficulties associated with reporting see IEEP (1993).

[45] In turn, EU citizens consider the Commission as the most appropriate recipient for complaints and advocate of EU environmental objectives. For more information on the Commission's relations with environmental NGOs see Webster (1998).

[46] According to a former DG XI secondment official (complaints department), interview, 10. May 1995, Glasgow.

[47] See *ENDS Report* (1995).

[48] During an interview, one Commission DG XI official defended the proposal, while his colleague opposed the proposal: 'We already have consultation procedures (...), the check could be counter-productive.' Brussels, 7. March 1997.

[49] Krämer (1997) describes the differences in legal cultures and their impact on EU environmental policy implementation. In a research interview, Ken Collins (at the time of the interview MEP and chairman of the EP Environment Committee) pointed out the differences in transposition approaches. He went so far as to describe the Italian attitude towards Directives as merely 'something to aim at'. Interview, 3. February 1995, East Kilbride.

[50] See Chapters 4, 5 and 6. Knill and Lenschow (1998) also argue that disappointing implementation outcomes cannot be explained with incompatible (national) administrations alone. Rather, a complex mix of factors shape policy implementation.

[51] Their discretion, of course, depends upon the form of EU environmental policy: Regulations have direct effect and are therefore transposed verbatim while Directives provide a large scope for interpretation and formal transposition.

[52] For instance, Baker et al (1994) argue that 'the weak resource base of peripheral regions

and Member States limits their capacity to implement policy. [This is] especially so in the field of environmental protection, requiring as it does high levels of scientific and technical expertise (...)'.
[53] Krämer (1997) highlights the lack of administrative staff and resources.
[54] For instance, it took the Member States ten years to agree on a common policy on environmental impact assessment. Once the EIA Directive (85/337/EEC) was adopted, implementors received no encouragement in putting the policy objective into practice and showed limited enthusiasm towards the Directive. See Chapter 6.
[55] Among others, Mann (1996) uses the term 'lowest common denominator'.
[56] Krämer (1997) and di Fabio (1998) use this description.
[57] Krämer (1997) would go as far as describing some of the opposition against 'imported' environmental provisions as 'xenophobic'.

References

Baker, S, Yearley, S., Milton, K (eds) (1994), *Protecting the Periphery: Environmental Policy in Peripheral Regions of the European Union*, Frank Cass, London.

Bakkenist, G. (1994), *Environmental Information. Law, Policy and Experience*, Cameron May, London.

Böhmer-Christiansen, S. and Skea, J. (1991), *Acid Politics: Environment and Energy Policies in Britain and Germany*, Belhaven Press, London.

Bomberg, E. and Peterson, J. (1999), *Decision-making in the European Union*, MacMillan, Basingstoke.

Butt Philip, A. (1994), *Regulating the Single European Market: A Comparison of the Implementation of Social and Environmental Legislation*, Research Paper.

Commission (1994), *Interim Review of Implementation of the European Community Programme of Policy and Action in Relation to the Environment and Sustainable Development. Towards Sustainability*, COM (94) 453 final.

Commission (1994a), *Annual Report to the European Parliament on the implementation of EU environmental policies*, OJ No.C 154/42.

Commission (1996), *Proposal for a European Parliament and Council Decision on the Review of the European Community Programme of Policy and Action in Relation to the Environment and Sustainable Development. Towards Sustainability*, COM (95) 647 final.

Commission (1996a), *Taking European Environmental Policy into the 21st Century*, pamphlet.

Commission (1996b), *How is the European Union protecting our Environment?*, pamphlet.

Commission (1999), *The Global Assessment of the Fifth Environmental Action Programme. Europe's Environment. What directions for the future?*, COM (99) 543 final.

di Fabio, U. (1998), 'Integratives Umweltrecht. Bestand, Ziele, Möglichkeiten', *Neue Zeitschrift für Verwaltungsrecht* Nr.4, pp.329-337.

Dinan, D. (1999), *Ever Closer Union. An Introduction to European Integration*, MacMillan Basingstoke.

Demmke, C. (1996), *Verfahrungsrechtliche und administrative Aspekte der Umsetzung von EG-Umweltpolitik*, European Institute of Public Administration.

ENDS Report (1995), 'European Environment Agency gets under way', No.240, pp.2023.

Europe (1997), Nr.6903 (N.S.), p.6.

Freestone, D. (1991), 'EC Environment Policy and Law', *Journal of Law and Society*, No.18(1).

Haigh, N. (1984), *EEC Environmental Policy and Britain. An Essay and a Handbook*, Environmental Data Services, London.

IEEP (1993), *The State of Reporting by the EC Commission in Fulfilment of Obligations contained in EC Environmental Legislation*, London.

Keating, M. and Hooghe, L. (1996), 'By-passing the nation state? Regions and the EU policy process', in J. Richardson (ed), *European Union Power and Policy-making*, Routledge, London, pp.216-229.

Knill, C. and Lenschow, A. (1998), 'Coping with Europe: The impact of British and German administrations on the implementation of EU environmental policy', *Journal of European Public Policy*, vol.5, No.4, pp.595-614.

Krämer, L. (1992 and 1997), *Focus on European Environmental Law*, Sweet and Maxwell, London.

Liberatore, A. (1991), 'Problems of transnational policymaking: Environmental policy in the EC', *European Journal of Political Research*, No.19, pp.281-305.

Mann, M. (1996), 'EU struggles to find right shade of green', *European Voice*, 30.May-5.June, p.13.

McHugh, F. (1996), 'Voluntary accord get mixed response', *European Voice*, 30.May-5.June, p.18.

Nugent, N. (1999), *The Government and Politics of the European Union*, MacMillan, Basingstoke.

Official Journal (1973), 'First Environmental Action Programme', No.C, 112/1.

Official Journal (1977), 'Second Environmental Action Programme', No.C, 139/1.

Official Journal (1983), 'Third Environmental Action Programme', No.C, 46/1.

Official Journal (1987), 'Fourth Environmental Action Programme', No.C, 328/1.

Official Journal (1993), 'Fifth Environmental Action Programme', No.C, 138/1.

Smith, M. (1997), 'Brussels in environmental clampdown', *Financial Times*, 18/19. October, p.2.

Webster, R. (1998), 'Environmental Collective Action. Stable patterns of co-operation and issue alliances at the European level', in J. Greenwood and M. Aspinwall, *Collective Action in the European Union. Interests and the new politics of associability*, Routledge, London, pp.176-195.

Werner, J. (1996), 'Das EU-Netzwerk für die Umsetzung und Vollzug des Umweltrechts', in G. Lübbe-Wolff (ed), *Der Vollzug des Europäischen Umweltrechts*, Erich Schmidt Verlag, Berlin, pp.131-138.

Wood, C. (1995), *Environmental Impact Assessment: A Comparative Review*, Longman, Essex.

World Commission on Environment and Development (1987), *Our Common Future*, (Brundtland Report).

Young, S. (1993), 'Environmental Politics and the European Community', *Politics Review*, vol.2(3), p.6.

4 Environmental Politics and Policy in the UK and Germany

Introduction

The previous Chapter described the complexity of the EU environmental policy area and highlighted common difficulties in *filtering* EU environmental policies through the implementation *layers*. It argued that implementation *layers* are diverse, complex and often incompatible with the style and content of EU environmental policies. This Chapter takes up the incompatibility argument by examining the national *layers* of the UK and Germany and their environmental policies in more detail. In particular, the Chapter examines the national environmental policies and how they have been shaped by distinctly British/ German formal and informal *determinants*. It then assesses to what extent national and EU environmental policies differ and highlights those formal and informal *determinants* in the UK and German *layers* which hinder the implementation of many EU environmental policies. With the national *layers* in mind, Chapter 5 can then proceed with an investigation of the *sub*national *layers*, Scotland and Bavaria, whose environmental policies and implementation performances in the EU environmental policy area heavily depend upon their embeddedness within the wider state systems.

Focusing on the national *layers*, Germany and the UK have often been described as environmental 'leader' and 'laggard' respectively.[1] Indeed, as far as EU environmental policy-making is concerned, Germany frequently has put pressure on other Member States to set uniform environmental standards, while the UK has blocked many EU environmental policies which it considered over-ambitious or unnecessary.[2] However, the 'leader-laggard' analyses miss one important point: in terms of EU environmental policy implementation, both Germany and the UK have failed to realise many policy goals. At the beginning of the 1990s, Germany and the UK failed even to notify the Commission about the formal transposition of Directives in 10% of the cases. This performance improved slightly by 1995.[3] Germany and the UK have failed to implement various types of environmental policies: the Commission has tackled both Member States over the inadequate implementation of the EIA Directive (a typical procedural framework Directive), the failure to designate areas for environmental protection (under the Wild Birds and Habitats Directives), and failure to comply with quantitative and qualitative environmental standards (in particular water and air quality standards).[4] Over the years, the Commission referred alleged infringement cases to the ECJ, which confirmed the Commission's criticism on a number of occasions.

The following investigation gets to the bottom of the implementation problems by examining the national *layers* and their environmental policies in detail. The Chapter argues that both Member States feature 'distinct' environmental policies which derive from their formal and informal circumstances (or *determinants*) and differ in many respects from EU environmental policies. The Chapter suggests two broad reasons for the often inadequate implementation of EU environmental policies:

- Firstly, EU environmental policies often clash with informal *determinants*, in particular with national policy-makers' priorities and strategies as well as policy styles and practices
- Secondly, EU environmental policies are often incompatible with formal *determinants* such as political-administrative structures and legal systems that organise and administrate environmental policies within the national *layers*.

The United Kingdom

The 'Forerunner' and 'Dirty Man'

National environmental policies are shaped by formal constitutional settings as well as informal circumstances (e.g. policy priorities and relationships between actors) within state systems (or *layers*). In the case of the UK, formal and informal *determinants* have contributed towards a paradoxical environmental policy. The UK has often been at the forefront of environmental policy, while at other times political and economic behaviour in the UK has pointed towards a lack of environmental commitment. It is therefore difficult to pin down one straightforward UK policy on the environment.[5]

One of the key formal *determinants* contributing to the

paradoxical handling of environmental matters has been the UK constitutional setting, which is highlighted in figure 4.1. Until the devolution process of 1999, the UK has been essentially a centralised state which evolved over centuries and comprised four former kingdoms. Ultimate sovereignty has been with Parliament. Further, the UK has never possessed a written constitution as such. Therefore legislation has derived either from traditional conventions or political decisions taken in parliament, with the House of Commons playing the central role in decision-making (although, since 1999 certain powers have been devolved to Cardiff and Edinburgh). UK policies have been formulated as Acts of Parliament, Regulations, and Statutory Instruments supported by administrative Circulars. Scotland and Northern Ireland have always required separate bills to allow for their legal traditions as well as geographical and other diversities, while England and Wales have shared one legal framework. Despite these variations, legislation for England, Wales and Scotland have stemmed from the legislative centre in Westminster and Whitehall and have not departed significantly from each other in terms of content and objective. This convergence, however, is in the process of changing in the light of devolution.

The combination of ancient traditions on the one hand and the potential for radical change in parliament on the other is reflected in the development of UK environmental policy. In some respects the UK has been slow in adopting a strong environmental position, while in other instances the UK has taken the lead in environmental initiatives.

The Forerunner

In terms of informal attitudes towards environmental protection, the British have traditionally valued their natural environment and initiated environmental organisations and legislation long before other states (and indeed the EU) even considered the problems of pollution. The Commons, Open Spaces and Footpaths Preservation Society, for instance, was founded in 1865 and was one of the first environmental NGOs to appreciate the countryside and campaign for its access. Modern environmental NGOs in the UK continue this tradition; today they enjoy large memberships and are among the wealthiest and most influential environmental groups in Europe.[6] Building on the early influence of environmental NGOs, the UK was also the first country in Europe to form a Green Party, which, however, could not establish itself as a strong parliamentary force mainly because of the UK electoral system (a formal *determinant* constraint).[7]

Apart from a traditional interest in the countryside, the British have also been environmental 'initiators' in Europe because they were the first to suffer under the negative effects of industrial activities which began in the UK. Industrialisation brought not only material prosperity but also problems of pollution. Legislation on the environment, dating back as early as 1273, responded to the negative effects caused by early industrial activity.[8] 1863 saw the Alkali Act, which established a framework to control industrial processes that emitted hydrochloric acid. One century later, the British were the first to respond to environmental pressures by adjusting their formal government structure: in 1970 they established a Department of the Environment (DoE).

The establishment of the DoE (today the Department of the Environment, Transport and the Regions, DETR) was facilitated by the UK policy style and in particular by the UK-specific definition of the term 'environment'. Under the environment, people in the UK have understood a collection of issues, which included architecture, town planning, local government administration and pollution control. The DoE was established to accommodate this wide range of policy matters. In this sense, the DoE concept resembles the EU environmental policy approach of policy 'integration'. However, environmental interests in the UK have tended to take a subordinate position in relation to other policy priorities such as housing and industrial development.[9] A similar (integrative but also compromising) pattern applied to the territorial Scottish Office (and now the Scottish Executive), which dealt with environmental policies in Scotland (outlined in detail in Chapter 5). EU environmental policy objectives have been processed accordingly throughout the UK: next to other (economic or social) interests, many EU environmental policies have lacked the support by UK actors necessary to ensure effective implementation.

The 'Dirty Man'[10]

To a certain extent, the UK has taken the environmental policy lead in Europe, for instance by establishing environmental institutions and organisations and adopting legislation on air and water quality. However, these initiatives have represented piecemeal and 'reactive' rather than comprehensive and precautionary measures to pressing environmental problems. While UK measures signalled a beginning in the environmental policy area, they have never merged into a coherent and consistent policy pattern. Several informal *determinants* relating to the UK policy style and practices have contributed to the half-hearted UK environmental policy.

Firstly, UK political-administrative actors have followed a rather pragmatic policy approach. This short-term, step-by-step approach has had the advantage that only realistic goals have been envisaged and that UK policy-makers have committed themselves only to policies which they could confidently implement. On the other hand, UK 'pragmatists' have tended to plan for the immediate future only, have been reluctant to take on board scientific uncertainties and vague predictions and have found it difficult to pursue 'unnecessary' and 'radical' policies just because of unsubstantiated worries and public demands. Since many environmental problems are difficult to measure and predict, environmental policies have tended not to fit into the UK pragmatist mould. As far as EU environmental policies such as the Directive on large combustion plants are concerned, UK actors have perceived many of their preventive and stringent standards as unnecessary burdens which overstep UK marks. 'Ambitious' EU environmental policies have therefore been pursued by UK actors with a certain reluctance or scepticism.

Not far from the UK 'pragmatism' lies the often praised 'impartiality ethos' of the UK civil service.[11] UK administrators have been described as committed professionals who implement legislation with a certain 'sense of neutrality'.[12] This professional neutrality should be advantageous for the pursuance of environmental objectives in general and EU environmental policies in particular. Yet, this 'neutrality' has had the effect that often the source of legislation, be it British or European, has been unknown. As a result, implementors could not clarify and coordinate policy objectives with the 'makers' of a policy. In addition, the impartiality ethos has been limited when it came to the implementation of environmental measures which UK administrators perceived as unnecessary or costly.[13] Therefore, the impartiality ethos may have positively influenced the implementation of some individual EU environmental policies, but has been more likely to evaporate in the majority of other policy cases.

Another explanation for the half-hearted UK environmental policy concerns the preference for voluntary action in environmental matters. UK Governments and Conservative Governments in particular, have supported the general view that any restrictions to economic prosperity should be avoided and that UK citizens as well as the private and public sectors should not be forced to pay for 'costly' environmental objectives. This emphasis on voluntary action, however, has secured only a relatively small number of 'green' initiatives from individual citizens,

NGOs, private companies and local authorities. As Young (1993) remarks, in the UK leaflets promoting environmental interests 'are everywhere'. These and other voluntary initiatives, however, have not provided for consistent and effective environmental action in the UK.

The policy approach of voluntary action together with the integrative definition of the environment is reflected in the UK legal system (a formal *determinant*) and the way in which UK legislators and administrators have dealt with environmental policies. More recent EU environmental policies which are based on voluntary action and policy integration (e.g. eco-audit and eco-labelling Regulations) have been more compatible with UK political-administrative structures and legal instrument preferences than earlier EU environmental policies which focused on quantitative standards and strict qualitative regulations. UK administrators have therefore found the implementation of more recent EU environmental policies comparatively easy.

With regard to legal instruments containing high quantitative or qualitative standards and preventive policy objectives, UK administrators and citizens have preferred to trust the self-healing potentials of the natural environment, i.e. the capability of nature to absorb diluted and dispersed polluting substances.[14] As Böhmer-Christiansen (1990) put it, soil, water and air have been seen as free resources for waste disposal:

> until the moment of unacceptable harm or damage is reached.

Despite an early interest in countryside issues, the overall UK policy style and approach towards environmental problems has been reactive rather than preventive. Action has been taken only when environmental damage became unbearable and when pollution incidents became potential threats to human health. As a result, many stringent and preventive EU environmental policies such as the Habitats Directive and water quality Directives have been pursued UK political-administrative actors only after considerable pressure from environmental interest groups and the EU Commission.

One key reason why 'green' considerations could not take central stage in UK politics can be seen in the political-administrative actors' inability (and/ or reluctance) to completely open long-established lobbying networks to include environmental NGOs. Therefore, traditional political-administrative structures and informal relationships between actors have played a significant role in the setting of political priorities. While countryside lobbyists and 'clients' from industry and the farming community have maintained strong links with Westminster and Whitehall, new environmental NGOs have been generally looked upon with mistrust and scepticism.[15] Only in recent years have environmental NGOs gained more access to political-administrative actors and are currently establishing themselves as influential campaigners and advisers on environmental matters in the UK.

In terms of parliamentary representation, figure 4.1 highlights a major formal constitutional obstacle: environmentalists have not been able to enter the House of Commons due to the electoral system (the first-past-the-post or FPTP-system) [16] which has tended to give weight to either of the two mainstream parties, the Labour Party and the Conservative Party. For this reason, the UK Green Party has never gained as much political influence as its European counterparts, although it was the first political party in Europe to specifically address environmental concerns. Meanwhile, the Labour Party, the Conservatives and the Liberal Democrats have responded to the campaigns of environmental NGOs and have adopted 'green' issues in their party manifestos.[17] The main parties have adjusted priorities and strategies slightly to accommodate environmental pressures. Minor changes have included, for instance, a few lines in party manifestos and a brief mentioning of environmental issues at party conferences. These changes have proven to be sufficient to satisfy the majority of party members and voters but have not provided for a strong environmental force in parliament which monitors the progress and effectiveness of EU environmental policies.[18]

Despite its forerunner position, the UK has pursued a half-hearted environmental policy, which has prompted observers to describe the UK as a 'dirty man'. Frequently, post-war UK governments have blocked the adoption of international and EU environmental policies and have given the impression that environmental issues take second place in policy priorities. For instance, the UK initially vetoed the 1987 EC Directive on large combustion plants, installed only seven nitrogen dioxide monitoring stations compared to 200 in Germany, and opposed Directives regulating water quality. The latter were considered as unnecessary by UK actors because of the UK's island situation and the self-healing potentials of its 'fast-flowing' rivers. Internationally, the UK joined the USA and Saudi Arabia to produce the worst records in preparation for the UN Conference on Environment and Development in Rio (1992), according to a consortium of more than 100 NGOs.[19] Further, the UK initially delayed the signing of the biodiversity convention and was reluctant to promote a new UN environmental body as a follow-up measure to the Rio Conference. Two

UK initiatives, the International Conference on the Ozone Layer in February 1989 and the inclusion of the environment on the G7 agenda in July 1989, could not counter the UK's image of a 'dirty man'.

UK Governments and the Environment

The half-hearted attitude towards environmental concerns is reflected in the policy priorities and strategies of consecutive UK Governments in the last decades. From 1979 to 1997, Conservative Governments fundamentally changed the political landscape in the UK by introducing a 'laissez-faire' economic policy, which included radical privatisation measures. In the process, environmental considerations were pushed to the sideline of UK politics. In their efforts to remove and avoid economic obstacles, Conservative Governments relied heavily on voluntary environmental action and tended to 'react' to pollution problems rather than innovate policy reforms.[20] Next to other policy issues, the environment was perceived a low-priority area. Taken together, the formal restructuring of UK government (culminating in the loss of many administrative powers to the private sector) and the voluntary action approach aggravated pollution control and other environmental measures. However, the privatisation process implied also positive consequences for the environment. Due to public scrutiny and (then) EC alertness the privatisation process of water and electricity could only be pursued in conjunction with the adoption of more stringent environmental measures.

Accommodating pressures from the (then) EC level and UK environmental NGOs for more environmental consideration, prime minister Thatcher (1979-1992) initiated a moderate shift in policy which was taken up by subsequent Conservative Governments and, indeed, the Labour Government under Tony Blair. Learning her lesson from an unpopular comment at the annual conference of the Scottish Conservative Party in May 1982, describing environmental issues as 'humdrum', she declared in 1988 during a speech to the Royal Society that protecting nature was:

> one of the great challenges of the late twentieth Century.

She further demonstrated her change in attitude by replacing the Secretary of State for the Environment, Nicholas Ridley, with the more sympathetic Chris Patten. Margaret Thatcher also created and chaired a Cabinet committee which prepared a White Paper on the environment. The 1990 White Paper titled *This Common Inheritance* signified the UK

Government's half-hearted approach towards environmental policies in so far as it contained a mix of fundamental as well as minor policy proposals. One of the White Paper's key policies was the energy efficiency policy which included a ministerial committee dealing with global warming. The Paper also included elaborate sections on issues such as protecting Cathedrals and abandoned shopping trolleys. As far as car traffic pollution was concerned, the Government suggested:

> adopting less aggressive driving habits to save fuel and keeping cars well tuned.[21]

Due to tensions between sectoral interests inside the Cabinet and Whitehall,[22] the White Paper turned out to be only a moderate Government policy on the environment. Yet despite the shortcomings, the White Paper represented a potential stepping-stone for more environmental policy commitments.

John Major followed in Margaret Thatcher's footsteps: environmental issues were neither at the top of the political agenda, nor could John Major completely ignore pressures for more environmental action. One major policy change occurred during his premiership with the introduction of 'green' fiscal instruments such as the landfill tax. Moreover, less ambitious road building plans towards the end of John Major's term in office, i.e. the reduction of road building expenditure from over £8 billion (3 year budget until 1996) to £6 billion (3 year budget until 1999),[23] indicated a response to increased traffic pollution and heightened public awareness.

Tony Blair's Labour Government has followed its predecessors' environmental policy line by taking up the fiscal policy as a convenient policy tool to tackle pollution.[24] Occasionally, the Labour Government has demonstrated a keen interest in environmental matters. For instance, in February 1998 the Labour Government proposed new housing development plans for 'brown field' sites instead of 'green field' sites, thereby avoiding further destruction of the countryside. In June 1998 the Labour Government published a white paper on transport, which proposes charges for the use of roads and parking. The charges should deter motorists from using their cars and encourage them instead to use public transport. However, other examples such as the Government's earlier decision in July 1997 not to oppose major road building projects have signalled that the Labour Government is not prepared to undertake radical environmental policy reforms in the near future. The Labour Government's initiatives have to date not merged into a coherent and effective environmental policy which takes account of sustainable development. Environmental considerations are not fully integrated into other (economic and social) policy areas and more recent Government initiatives do not signal the beginning of a new stringent UK environmental policy that would pave the way for EU environmental policies in future. It is therefore not surprising that the Labour Government's environmental policy has been described by environmentalists and the media alike as 'pale green'.[25]

In sum, a combination of formal and informal *determinants* has influenced the development and conduct of environmental policy in the UK. Environmental matters have been processed in a pragmatic manner, which has tended to exclude long-term, less tangible EU environmental policy objectives. In addition, UK policy-makers and administrators have focused on policies which react to (rather than prevent) environmental problems. They have relied upon voluntary environmental action and have been reluctant to open traditional lobbying networks to include new, 'green' lobby groups. At the parliamentary level, environmental interests have received limited support from politicians. On the other hand, a traditional interest in the countryside and early problems with industrialisation encouraged UK citizens to become forerunners in a number of environmental policies and organisations. While the UK has paved the way in some respects for other states and the EU, UK citizens and politicians have not utilised their forerunner position to establish a 'model' environmental policy which ensures sustainable development. Instead, the UK has suffered from the image of the 'dirty man of Europe, which was aggravated by the Conservative Governments' 'laissez-faire' economic policy. Despite attempts by Tony Blair's Government to improve the UK environmental policy, more commitment will be required to eliminate the 'dirty man' image in the near future.

Figure 4.1: The UK Layer and Environmental Policy

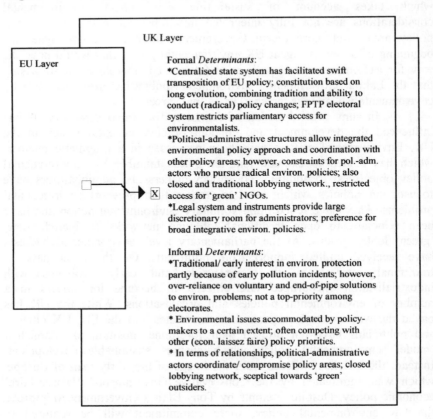

UK Layer

EU Layer

Formal *Determinants*:
*Centralised state system has facilitated swift transposition of EU policy; constitution based on long evolution, combining tradition and ability to conduct (radical) policy changes; FPTP electoral system restricts parliamentary access for environmentalists.
*Political-administrative structures allow integrated environmental policy approach and coordination with other policy areas; however, constraints for pol.-adm. actors who pursue radical environ. policies; also closed and traditional lobbying network., restricted access for 'green' NGOs.
*Legal system and instruments provide large discretionary room for administrators; preference for broad integrative environ. policies.

Informal *Determinants*:
*Traditional/ early interest in environ. protection partly because of early pollution incidents; however, over-reliance on voluntary and end-of-pipe solutions to environ. problems; not a top-priority among electorates.
* Environmental issues accommodated by policy-makers to a certain extent; often competing with other (econ. laissez faire) policy priorities.
* In terms of relationships, political-administrative actors coordinate/ compromise policy areas; closed lobbying network, sceptical towards 'green' outsiders.

The Federal Republic of Germany

The 'Latecomer' and 'Green Man'

Environmental policies in Germany, too, have been shaped by the formal constitutional setting as well as informal circumstances within the national *layer*. The Federal Republic of Germany has often been described as a state with an activist attitude (and policy) towards the environment.[26] However, in comparison with the UK, the interest in pollution problems has been a relatively recent phenomenon in Germany. This is partly due to the fact that the German industrial revolution commenced after British industrialisation and that environmental problems became noticeable in Germany some time after the first pollution incidents in the UK. Other explanations for the comparatively recent, but at the same time more rigorous, 'green' approach in Germany point towards other formal and informal *determinants* which are listed in figure 4.2.

The Latecomer

In comparison with UK environmental legislation, which dates back to the 13th Century, the first German measure to control pollution was introduced in 1845 with the Prussian General Trade Ordinance (Gewerbeordnung).[27] The late adoption of the environmental issue was not only due to Germany's comparatively late industrialisation, but also due the ongoing dispute over formal competencies between the Reich Government and Confederal States which prompted polluters to take environmental measures in their own hands. Facing severe river pollution, representatives from the agricultural and industrial sectors had a commercial interest in clean water and established common water quality standards which would apply throughout the German Reich.[28]

   In the first half of this Century, environmental matters did not receive much attention due to National Socialism and two World Wars which dominated and devastated Germany. Only after the alarming side effects of the 1950s' Federal German economic miracle (Wirtschaftswunder) became apparent, did citizens notice environmental problems. Again, initiatives to combat pollution came mainly from the Länder level, predominantly from North Rhine Westphalia which suffered most under the side effects of industrial activities. The pollution issue was eventually taken up at the Federal level: the SPD under Willy Brandt campaigned for 'blue skies over the Ruhr' during its 1961 Bundestag election campaign.[29]

Obviously, one of the key formal reasons behind the slow adoption of an environmental policy has been the traditional separation of government levels, which has often resulted in disputes over competencies and has hindered the progress of many environmental policies including EU environmental policies. The separation of political-administrative competencies continues to this day: established in 1949, the Federal Republic of Germany is based on a written constitution which provides for checks and balances between the executive, the judiciary and the legislature as well as Federal and Länder government levels. The Federal Parliament (Bundestag) shares decision-making powers with the Länder, which are represented in the regional chamber (Bundesrat). In addition to the balancing of Länder and Federal competencies, judicial review can scrutinise policies adopted in Germany. Federal as well as Länder legislation is adopted in the form of Acts of Parliament (Gesetze), Regulations (Rechtsverordnungen), and Administrative Instructions (Verwaltungsvorschriften).

Focusing on formal political-administrative structures, the German constitution manifests and emphasises the vertical separation of competencies as well the sharing of powers between government levels. Consequently, political-administrative actors at *sub*national and local levels have taken a great interest in protecting their areas of autonomy. In other words, formal conditions have had an impact on informal perceptions. Apart from protecting their autonomy in decision-making, *sub*national and local implementors have tended not to report to government levels above on policy performances.[30] For the implementation of EU environmental policies, this lack of communication has been particularly unfavourable: since most of them have required implementation on the ground, EU policy-makers as well as national (and *sub*national) implementors have been left in the dark over the effectiveness of policies.

Apart from cutting across government levels, EU environmental policies have tended to cut across policy sectors which, in the case of Germany, have enjoyed considerable independence from each other. In comparison with UK horizontal structures and institutions, German environmental institutions and agencies in particular have been able to pursue ambitious policies without compromising environmental objectives and giving way to immediate pressures from other sectors. On the other hand, the independence of sectoral institutions has resulted in a lack of communication and cooperation. In the case of environmental policies, this gap has hindered the widespread adoption and integration of environmental objectives into other policy areas. Yet many EU environmental policies such as the IPC and EIA Directives explicitly promote the integration of environmental considerations into other policy areas. Faced with the fragmented nature of German political-administrative structures, their implementation has been cumbersome and often disappointing.

In order to function properly, the fragmented and compartmentalised German political-administrative system has had to rely on a policy style which is based on consensus and conciliation of policy sectors and government levels. This conciliatory approach has not only prevented political paralysis, it also has ensured that policies are more acceptable for a wide spectrum of actors. On the other hand, the search for consensus has caused decision-making and the subsequent implementation of policies to be slow. Environmental matters in particular, which tend to affect other policy areas such as transport, have had to be processed through complicated consensus mechanisms. Environmental policies from the EU level, too, have had to face the scrutiny of 'affected' actors. The perceived right to assess and approve (or reject) every policy has often delayed the implementation of EU environmental policies. Moreover, German implementors have tended to re-shape 'policies from Brussels' to suit their policy preferences. Consequently, the implementation of many EU environmental policies has been delayed in Germany or failed to reach policy *targets*, according to Commission officials and environmental NGOs.[31]

The Green Man

Despite the fact that the German political system was relatively slow in addressing environmental problems, a mix of long-established formal and informal *determinants*, highlighted in figure 4.2, contributed to the establishment of a comparatively rigorous environmental policy. Firstly, as part of wider societal changes, attitudes towards environmental protection were influenced significantly by the 'new social movements' (in German Neue Soziale Bewegungen) of the 1960s, 1970s and 1980s.[32] NSMs raised environmental concern and put green issues on the agenda of politicians and the media. The new emphasis on environmental protection was facilitated further by the German electoral system[33] (a formal *determinant*) which helped the German Green Party (Die Grünen) enter the Bundestag in 1983 for the first time with 27 MPs.[34] Their success not only shocked the main parties, the Christian-Democratic Union (CDU) and the Social-Democratic Party (SPD), it also threatened the very existence of the small

FDP. While the three parties adjusted their party manifestos accordingly, Die Grünen changed the political landscape in Germany considerably, forming 'red-green' Länder governments with the SPD in Hesse, Berlin, Lower Saxony and North Rhine Westphalia. In October 1998, Die Grünen even succeeded in entering a 'red-green' coalition government with the SPD at the Federal level and replacing the Christian-Liberal Government under Helmut Kohl. Overall, environmental objectives have enjoyed comparatively strong public and parliamentary support, which has ultimately contributed towards some radical environmental policies inside Germany and influenced the EU in adopting some stringent EU environmental policies such as the Directive on large combustion plants.[35]

The German policy style and practices contributed towards a more rigorous environmental policy. In particular, the German definition of the term 'environment' (in German Umwelt) has meant that environmental objectives have been pursued in a 'concentrated' manner. However, while the apparent separation between environmental and other policy areas has facilitated the adoption of a number of far-reaching environmental policies, the environmental policy as a whole has suffered because green issues have often been considered on their own without taking into account wider contexts and other (economic) interests. This non-holistic perspective has resulted in many environmental policies having only limited impact. Nevertheless, the discretionary room and 'creativity' of environmental political-administrative actors has also meant that some far-reaching German policies have been models for EU environmental policies such as the comparatively early introduction of catalytic converters for cars and lead-free petrol.

More radical environmental policies have been adopted not only because of the discretionary room the environmental policy sector has enjoyed. Radical policies have also responded to, what Böhmer-Christiansen and Skea (1991) would call, 'Angst' (anxiety) over environmental threats. Concerns over pollution and environmental deterioration have been more intense in densely populated Germany than in many other European states. Environmental problems such as the Chernobyl accident or the much-publicised Waldsterben ('dying forests') hit a raw nerve with citizens in Germany, leading to vociferous calls for green policies and increased pressures on political-administrative actors to act. This Angst has influenced the development of environmental policies in so far as many policy objectives have been more substantial and far-reaching than in other EU Member States (and, in fact, the UK). Yet, Germany's preparedness to adopt stringent environmental standards without scientific backing has not resulted in a comprehensive policy that would substantially change behavioural patterns in Germany towards sustainable development. Nevertheless, the German Angst has had the advantage that EU environmental policies have not faced obstacles of acceptance, as has been the case in the UK.

In order to tackle environmental problems and perceived threats, Germany has focused on legal instruments which specify regulatory and (subsidised) technological policy solutions. The promotion of the 'state of the art' (in German: Stand der Technik) and the regulatory approach continue to dominate the German environmental policy, although in recent years Germans have warmed to the idea of voluntary environmental action. The techno-centric and regulatory approaches are most evident in the nuclear sector where citizens have been reassured for decades that nuclear accidents are impossible under strict regulations and modern technological-scientific management. Over the years, political-administrative actors and experts have developed an expertise and a certain perfectionism in applying and monitoring 'end-of-pipe' technologies. On the other hand, German political-administrative actors have been reluctant to alter their perspective in favour of a less technology-orientated approach. More importantly, they have found it difficult to accept EU environmental standards which depart from their own standards. As a result, many EU qualitative and quantitative requirements (such as water quality requirements) have not been implemented properly because of technical discrepancies. In addition, German implementors have taken the liberty to fill, what they perceived as, legislative gaps with technical and legalistic details where EU Directives are silent. However, by doing so, German implementors have tended to change the actual character of many EU policies. In the case of the EIA Directive, this 'perfectionism' has culminated in the criticism that German legislators over-shot their marks: in two legal cases (C431/92 and C396/92) the ECJ pointed out that the Federal German specification of project categories was detailed but failed to mention many important project types which effectively excluded them from legislation.[36]

Despite this and other disputes, Germany has presented itself nationally and internationally as an advocate of high green standards, often much to the annoyance of other EU Member States.[37] However, while politicians in Germany have promoted a 'green advocate' image and have introduced some of the most stringent environmental measures, many environmental policies have originated from economic considerations or have been a response to obvious environmental problems and public pressure. Germany has not been the environmental 'Musterknabe' (paragon) as portrayed by many politicians. To illustrate the point, the late establishment of the Federal Ministry for the Environment (in German: Bundesministerium für Umwelt, Naturschutz und Reaktorsicherheit, BMU) was not the result of a well-planned and fully committed adjustment of formal structures to tackle environmental problems. Rather, its creation was a panic reaction to a major environmental crisis (the 1986 Chernobyl nuclear accident), which subsequently caused a political crisis (the Federal Government was criticised over its inadequate handling of the accident).

Stringent German environmental legislation and standards have not concealed the fact that Germany has been one of the worst polluting countries in the world.[38] Moreover, the economic strains of the 1990s have put a damper on the environmental policy, despite the political success of Die Grünen and NSMs. Particularly, German Unification and obligations associated with the Economic and Monetary Union (EMU)[39] have pushed environmental considerations to the sideline of German politics. Economic and social policy objectives are now at the top of the German priority list with the effect that many environmental policy adjustments of the 1970s and 1980s are watered-down and many EU environmental policy objectives are neglected to eliminate any obstacles against the 'Wirtschaftsstandort Deutschland'.[40] Despite fiscal and other reforms which include an environmental agenda, it still remains to be seen to what extent the 'red-green' coalition government under Gerhard Schröder will steer German politics back towards a (truly) 'green man' position. Considering that Federal Chancellor Schröder is adamant to further boost German economic confidence, it is unlikely that Germany will adopt and implement an effective policy that would ensure sustainable development in Germany.

Federal German Governments and the Environment

Former Federal Chancellor Helmut Kohl and his Government colleagues were at the forefront of portraying the Federal Republic of Germany as a green state. Following a Social-Liberal Government initiative, Christian-Liberal Governments adopted the environmental principles of the 1971 Environmental Programme[41] and took up environmental policy ideas which had been in the pipeline at the time of government change-over in 1982. The large combustion plants legislation, in particular, signalled a start in the Christian-Liberal environmental policy and was a response to the alarming media coverage of the 'dying forests'. While Federal German Governments under Kohl pursued some of the most ambitious environmental policies, their environmental commitment was soon questioned by the public. Interior Minister Zimmermann's mishandling of the Chernobyl nuclear accident in particular highlighted the Governments' inability to cope with major environmental problems.[42] Walter Wallmann, the first Minister for the Environment leading the BMU, was new to the environmental policy area and was soon criticised over his handling of incidents such as the Sandoz fire.[43] From 1987 onwards, Wallmann's successor, Klaus Töpfer, assumed a more pro-active approach. However, Töpfer's symbolic commitment towards the environmental cause[44] could not disguise the low priority stance of his Ministry, which received only 0.3% of the Federal Budget.[45] Nevertheless, while Töpfer was in office, the Federal Government established a number of green policies such as the plastic bottle policy (a recycling system similar to the recycling policy in Denmark) and the Grüne Punkt recycling scheme for packaging waste of consumer goods.[46] The Grüne Punkt policy in particular encouraged the EU to consider similar policies on waste and packaging.

German Governments have established some of the most stringent environmental policies in Europe. At the same time, these measures have fallen short of a coherent environmental policy which controls the negative impacts of the German 'power house'. Moreover, German Governments have tended to react to public pressure rather than initiate policies which ensure sustainable development in Germany. Critics of the Kohl Government (in office until October 1998) would argue that environmental principles existed on paper but not in practice. For instance, many policies relied on a technical 'end-of-pipe' strategy which runs counter the principle of pollution prevention. Also, in the light of an ever-growing mountain of waste in Germany, polluters have obviously not paid the price for environmental damage despite the introduction of the Grüne Punkt and other recycling schemes. Although Germany possesses some of the most stringent environmental standards, the environment itself continues to deteriorate.[47] In the light of German Unification and European Integration, the Kohl Government focused on other policy priorities such as transport. Road and air traffic developments already overstepped forecast marks prior to Unification.[48] Despite an alarming increase in traffic and its pollution impacts, the Kohl Government committed itself to several large-scale projects intended to connect infrastructures in Eastern and Western Europe. Investments on public transport technology could not counterbalance the fact that the current Infrastructure Plan (until 2012) has been the most ambitious plan since the end of the Second World War.[49]

As coalition partners of the current 'red-green' coalition government, Die Grünen have tried to press for more radical environmental policies, however with limited success. Among other constraints, Die Grünen have had to put up with a less enthusiastic coalition partner and, more generally, with a fragmented political-administrative system that contributes towards a slow decision-making process. In this system, decisions are largely based on horizontal and vertical consensus which hinders self-initiative and leadership. But one of the main obstacles that has hindered a 'truly green' policy has been the resistance from certain parts of the business community in Germany and some of the Länder. For instance, the 'red-green' Government's policy to phase-out nuclear energy in Germany has faced severe obstacles from the nuclear sector lobby and (notably) the Bavaria State Government. The nuclear sector refuses to cooperate and threatens to re-locate to neighbouring Eastern European countries. The Government has so far failed to completely resolve the problem of nuclear accident threats.

In sum, Germany features a complex and to a certain extent ambitious environmental policy. Environmental matters have received much informal public and parliamentary attention due to the perceived environmental threats and the support from NSMs and the German electoral system. Yet, the formal constitutional setting and political-administrative structures have contributed to the comparatively slow progress in adopting and, more importantly, implementing environmental policies. In addition, political-administrative structures have been inflexible to policy priority changes as well as 'instructions from outside' (in particular from the EU). Horizontal and vertical gaps in the political-administrative structures have provided environmental actors with considerable discretion and non-interference from other sectors. The German 'compartmentalism' has also been detrimental to environmental policy integration into other policy sectors, which has been one of the main objectives of EU environmental policy.

German policy-makers have accepted and adopted some ambitious environmental objectives without necessarily relying upon scientific evidence. However, this preparedness to go beyond scientific proof has not always guaranteed the policies' successful implementation at a later stage. German technological expertise has been beneficial for the pursuance of some EU environmental objectives (such as installing NO2 monitoring stations). Yet, this over-reliance on technological solutions has neglected other, non-technical solutions to environmental problems. Overall, Germany has not fully lived up to its image of a 'truly green' state despite the more recent involvement of Die Grünen in the Schröder Federal Government. In many respects the economic pressures of the 1990s have put a damper on the Germany's commitment towards national and EU environmental policies. As Malunat (1994) observed:

> the [environmental] situation [in Germany] has improved only marginally, in many respects it has deteriorated even more.

Figure 4.2: The German Layer and Environmental Policy

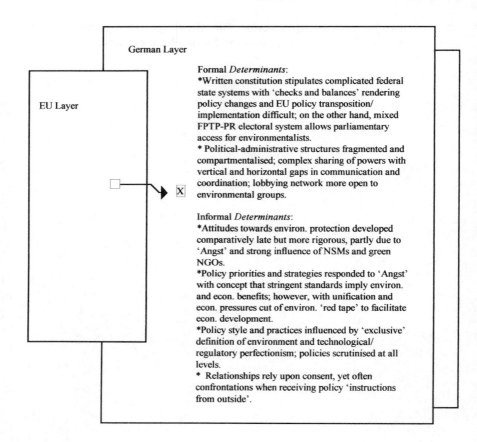

German Layer

EU Layer

X

Formal *Determinants*:
*Written constitution stipulates complicated federal state systems with 'checks and balances' rendering policy changes and EU policy transposition/ implementation difficult; on the other hand, mixed FPTP-PR electoral system allows parliamentary access for environmentalists.
* Political-administrative structures fragmented and compartmentalised; complex sharing of powers with vertical and horizontal gaps in communication and coordination; lobbying network more open to environmental groups.

Informal *Determinants*:
*Attitudes towards environ. protection developed comparatively late but more rigorous, partly due to 'Angst' and strong influence of NSMs and green NGOs.
*Policy priorities and strategies responded to 'Angst' with concept that stringent standards imply environ. and econ. benefits; however, with unification and econ. pressures cut of environ. 'red tape' to facilitate econ. development.
*Policy style and practices influenced by 'exclusive' definition of environment and technological/ regulatory perfectionism; policies scrutinised at all levels.
* Relationships rely upon consent, yet often confrontations when receiving policy 'instructions from outside'.

Conclusion: An Assessment and comparison of EU and National Layers

With the help of the *multi-layered implementation map* and particularly the *determinants*, the Chapter has described how two (EU member) states have developed and pursued their own environmental policies and how these policies have often been incompatible with EU environmental policies. The first finding therefore emphasises the discrepancy between national and EU *layers*.

(1) National layers have developed and pursued distinct environmental policies which have not always been in line with EU environmental policies.

It is true to say that the EU, UK and Germany have shared some common ground in the environmental policy area. Actors in the EU and national *layers* have acknowledged the necessity to adopt and implement environmental policies. They have been committed to more or less the same environmental principles (for instance, the precautionary and polluter pays principles) and have sought to accommodate environmental interests in a society, which is orientated towards economic and material prosperity. Secondly, actors in EU and national *layers* (national government ministers and representatives, Commission officials, EU and national experts etc.) have participated in the EU environmental policy-making process and have sought to influence policy decisions as much as possible. In this respect, EU environmental policies must have been, at least to a certain extent, compatible with Member States and their environmental policies. However, the complex and diverse dimension of the EU has also contributed an 'outside' impetus to national policies (i.e. they Europeanised national policies). While Germany and the UK have adapted to EU pressures, the European impetus has also brought elements 'foreign' to the national *layers*, which were ultimately incompatible with formal and informal conditions. As a result, many EU environmental policies have faced various formal and informal obstacles inside the national *layers* which have hindered the implementation of EU environmental policies.

By and large, the UK and Germany have featured their own, distinct, environmental policies which have depended upon, and have been shaped by, complex and unique mixes of formal and informal *determinants* inside the national *layers*.[50] EU environmental policies have therefore faced a number of informal and formal obstacles which have occurred in the form of different policy priorities, strategies and policy styles (i.e. informal *determinants*), as well as incompatible political-administrative structures and legal systems (i.e. formal *determinants*). These *determinants* have made the implementation of EU environmental policies difficult.

(2) EU environmental policies have often clashed with informal determinants such as policy-makers' priorities and strategies as well as policy styles and practices inside the national layers.

Actors in the EU and national *layers* have been involved in the balancing of informal policy priorities and strategies which has required the coordination of essentially two (conflicting) interests: environmental protection and economic prosperity. However, while all three *layers* have sought to consolidate environmental and economic interests, the input of economic considerations has differed considerably between the *layers*. At the EU level, the 'level-playing-field' consideration has played a significant role in the production of EU environmental policies. In contrast, UK (Conservative) Governments, have pursued a laissez-faire economic policy which has by and large pushed the regulation of environmental standards to the sideline of priorities. Since the early 1990s (and until October 1998), Federal German Governments under Helmut Kohl focused on the 'Wirtschaftsstandort Deutschland' policy which put a damper on many environmental priorities. This economic priority has not changed substantially with the 'red green' coalition government under Gerhard Schröder. In the light of these (varying) national economic strategies, many EU environmental policies, which were originally intended to harmonise and regulate 'polluting' economic activities, have encountered resistance from national actors and have therefore failed to reach their implementation *target*.

The Chapter has also highlighted the differences in policy styles and practices between the EU and national *layers*. Broadly speaking, UK actors have approached environmental matters with pragmatism, while their German counterparts have often pursued ambitious policies based on the precautionary principle and technological solutions. EU actors, for their part, have been influenced by a mix of Member States' policy styles and have developed their own 'European' style which has seen adjustments over the years. Since the early 1990s, EU actors and the Commission in particular, have tried to bring the EU and national *layers* closer in the environmental policy area by introducing a policy style which strengthens the links between EU and national actors via 'partnership' initiatives and dialogue groups. In addition, the Commission has pursued a less regulatory environmental policy which allows for more flexibility and subsidiarity. Whether the policies of partnership and subsidiarity can help overcome the dimensional hurdle from EU to national *layers* remains to be seen.

The UK and Germany have seen some 'Europeanisation' in their national environmental policies, i.e. they have given way to EU standards and objectives. Yet, there still remain formal and informal differences between EU and national *layers* which make EU environmental policy implementation difficult. Considering the differences between the *layers* in environmental policy, it is unlikely that the gap between EU environmental policy-making and implementation can be closed completely unless the EU environmental policy becomes an integral (and exclusive) part of the Member States' environmental policies.

(3) National layers' formal determinants, specifically their constitutional settings, political-administrative structures and legal systems, have often been incompatible with EU environmental policies.

With reference to formal structures, it is striking how the policy process in Germany has involved an enormously complex and complicated sharing of competencies.[51] The preparation, adoption and implementation of environmental policies within the German *layer* have therefore been a slow and arduous venture. Considering the complexity of German political-administrative structures, the difficulties associated with the implementation of (and compliance with) EU environmental policies come to no surprise. Actors at both Federal and Länder levels have felt obliged to add their own ideas to EU environmental policies and have adjusted EU policies to suit their national (and *sub*national) priorities and circumstances (see also Chapter 5 and case study in Chapter 6). According to the Commission and environmental NGOs, these 'adjustments' have in many cases contradicted with the original *targets* of EU environmental policies.

Until devolution in 1999, the UK has had an advantage over Germany in that national (and *sub*national) implementors were used to processing policies from central government in Westminster and Whitehall. In other words, EU environmental policies have not faced the same scrutiny by legislators and implementors in the UK, as has been the case in Germany (see also Chapters 5 and 6). However, this situation has changed recently and there are indications that actors in the devolved UK regions are not only processing EU policies separately from the centre, but also scrutinise them more closely and carefully. To date, however, this development has not (yet) let to substantially different transpositions of EU environmental legislation in England, Wales, Northern Ireland and Scotland, nor has it contributed towards a 'confrontational' attitude similar to that of the German Länder.

When it comes to the practical implementation of EU environmental policies, UK political-administrative actors have enjoyed considerable discretionary powers. This discretionary room has let to instances where the UK has not properly followed its EU environmental policy obligations. For example, UK officials considered the instalment of only seven NO_2 monitoring stations sufficient for the whole UK. This small number, however, prompted criticism from the Commission over the UK's inadequate implementation of an air quality Directive. Also, measuring water quality standards only in a small number of UK coastal areas not affected by water pollution attracted the Commission's attention. Therefore, the UK has not been spared from criticism over the inadequate processing of EU environmental policies.[52]

In terms of legal systems and instruments, both the UK and Germany have demonstrated some advantages as well as difficulties with EU environmental policies. Actors in Germany have often complained about the lack of clarity which characterises many EU environmental Directives. German actors have tended to fill legal gaps with their own criteria and have thereby changed the character of many EU policies. Where EU environmental policies have provided detailed information on qualitative and quantitative environmental standards, they have faced a different problem: often EU and German standards have been divergent and German actors have been reluctant, or unable, to make adjustments. The German environmental policy has been based predominantly on regulations and technological solutions to pollution problems which follow a long and elaborate decision-making process. EU legislation 'from outside Germany' has therefore been perceived as unnecessary and inconvenient unless they were in line with German standards (this even includes EU legislation which followed Federal German government initiatives).[53]

The UK, on the other hand, has preferred policies which provide discretionary room for implementors and focus on voluntary environmental action. More recent EU environmental policies such as the eco-audit Directive have therefore been more acceptable for UK implementors. UK actors have also followed a more 'integrative' approach towards environmental matters (demonstrated by the DoE/ DETR and, for Scotland, the Scottish Office/ Scottish Executive which consider environmental matters very much in conjunction with other policy sectors).[54] EU policies such as the IPC Directive have therefore been easier to implement in the UK than in Germany. This is an interesting observation, considering that the UK was initially opposed to the IPC policy. Germany, on the other hand, has had considerable difficulties in pressing the IPC policy into a fragmented, compartmentalised political-administrative system.[55]

In the case of qualitative and quantitative standards, however, UK administrators have often abandoned their 'sense of neutrality' when faced with EU policies which they considered too ambitious and stringent (i.e. policies over-stepping scientific marks and entailing 'excessive' costs). Moreover, UK actors have pursued quantitative and qualitative standards with less technological and regulatory perfectionism than German actors. In this respect, Germany has had an advantage over the UK.

Both German and UK actors have sought to influence EU policy-making as much as possible to carry through their environmental policy ideas. Both Member States have also dealt with the subsequent EU environmental policy implementation in accordance with their informal preferences and formal conditions. UK actors have reluctantly accepted EU Directives on air and water quality (many of which still require proper implementation), while German actors have implemented with, great difficulty, EU Directives which contain broad, integrative and procedural policy objectives.[56] In this respect, both national *layers* have demonstrated neither a 'euro-sceptic' nor 'euro-enthusiastic' attitude towards EU environmental policies. Another finding common to both *layers* has been a missing coherent and consistent overview on implementation performances, which in turn has contributed towards a lack of discipline. Conscious of the EU diversity and the difficulty in gaining an overview, Member States' implementors have tended to be lenient with EU environmental policy objectives, especially with objectives which affected other (economic) policy priorities. Even if all Member States exercised a certain amount of implementation discipline, mistrust over other Member States' lax implementation practices still remains.

The success of EU environmental policies has obviously depended upon the formal and informal *determinants* within the implementation *layers* (confirming Argument 1). This Chapter has provided a general picture of EU and national environmental policies from a wider 'domestic' (i.e. national *layer*) perspective. Yet, a 'domestic' perspective limited to the national *layer* is not sufficient for fully covering the actual ground where the bulk of EU environmental policies are implemented. Most EU environmental policies have to be *filtered* through to the regions and local communities and affect areas of *sub*national and local government competency such as planning, public health, waste and water management.[57] It is within the *sub*national *layers* that the majority of EU environmental policies are implemented or fail to be implemented. *Sub*national regions process policies within their own 'distinct' frameworks (or *layers*), which have to be taken into account when considering the EU environmental policy deficit. The following Chapter therefore addresses the *sub*national dimension by examining Scotland and Bavaria and their influence on EU environmental policies. It investigates to what extent Scotland and Bavaria confirm the above described characteristics of the national *layers*, or modify national *determinants* to suit their *sub*national circumstances, or to what extent they defy national *determinants* in order to present their own 'distinct' environmental policy (testing Argument 2).

Notes

[1] Sbragia (1996) identifies Germany as a 'leader' and the UK as a 'laggard' in EU policy-making.

[2] For instance, in the early 1980s, Germany took the lead in adopting Directives that regulate emissions from large combustion plants.

[3] For details see Commission (1996).

[4] Among others, Directives 85/203/EEC (air quality); 76/160/EEC (bathing waters); 80/778/EEC (drinking water); 79/409/EEC (wild birds); 92/43/EEC (habitats). For more information see, among others, Demmke (1996). For a British account on the implementation of water quality Directives see Ward et al (1995).

[5] McCormick (1991) describes the British attitude towards environmental matters as rather 'curious'.

[6] For a comparison see Gordon (1994).

[7] In the first-past-the-post electoral system (or FPTP-system) the candidate who secures more votes than his/ her rivals wins the constituency seat. The UK Green Party was founded in 1973 and was named Ecology Party between 1975 and 1985.

[8] Early environmental legislation included a law prohibiting the burning of sea coal.

[9] To illustrate the point, only 10% of the DoE/ DETR staff has dealt with environmental matters, according to McCormick (1993).

[10] Young (1993) describes Britain as a 'dirty man'.

[11] Siedentopf, H and Ziller (1988) describe the UK 'impartiality ethos'.

[12] Siedentopf (1990) uses the term 'Neutralitätsverständnis'.

[13] The case study in Chapter 6 highlights the critical views Scottish Office officials have had towards the EIA Directive.

[14] For example, UK actors have argued that fast-flowing rivers and the surrounding sea can absorb polluted water and that water quality legislation is therefore not necessary.

[15] Gordon (1994) contrasts UK and German attitudes towards 'outsiders': at an environmental conference, John Gummer (then Secretary of State for the Environment) refused to answer a question from the floor on the future of the THORP nuclear fuel

processing plant. In contrast to his German colleague Klaus Töpfer, who did not shy away from controversial questions, Gummer argued that the decision concerning THORP was for him alone and therefore not a matter for public consultation.

[16] With devolution, the Scots introduced a mixed FPTP-PR electoral system which has resulted in the first election of a Green (list-) member of parliament: Robin Harper MSP.

[17] The main parties also responded to the 1989 European Parliament election result. The UK Greens gained 15% of the votes cast, however they did not win any EP seats because of the (then) FPTP-system.

[18] For a detailed account see Robinson (1995). See also Maloney and Jordan (1995) who point out that in relation to other issues, only 4% of British adults nominate 'green' concerns as most important.

[19] See McCormick (1993).

[20] Blowers (1987) elaborates on the conflict between the Thatcher philosophy of deregulation and privatisation on the one hand and environmental problems on the other.

[21] Moreover, the Government considered it necessary to include the following statement in the White Paper on the environment: 'The Government welcomes the continuing widening of car ownership as an important aspect of personal freedom and choice. The speed and flexibility of motoring make it indispensable for much business travel, which in turn is vital for the economy.'

[22] In particular, transport and finance interests clashed with environmental objectives of earlier White Paper drafts. In 1998, John Prescott faced similar problems with his White Paper on the environment which was delayed due to internal disagreements on policy priorities.

[23] See *The Economist* (1997).

[24] Critics point out that eco-taxes are used as another source of public income and not so much as a tool to encourage 'green' behaviour. Eco-tax examples include the landfill tax which has been increased recently by 10% (from £11 to £12 a tonne of dumped rubbish) (*The Guardian* 2001) and a tax on water pollution for firms (*The Independent* 1997).

[25] Chancellor Gordon Brown's 1998 budget is described as 'pale green' in *The Economist* (1998).

[26] Among others, Sbragia (1996) describes the Germany's activist position/ image in EU environmental policy-making. Heritier et al (1994) evaluate (and qualify) the conventional perception that Germany is an activist state in environmental policy.

[27] The environmental measures were part of a general check on trade practices and were considered within the local authorities' discretionary powers. For more information see Weale (1991).

[28] Water control initiatives such as the 'Emscher Genossenschaft' were set up by local authorities and industry.

[29] 'Blauer Himmel über der Ruhr' cited in Malunat (1994).

[30] Strictly speaking, they are not required to report to superior levels. For instance, when asked about further details concerning the practical implementation of the EIA Directive, the Bavarian Ministry for the Environment did not have information from regional/ local authorities. Written correspondence, 7. January 1997.

[31] See for instance ECJ Case C-396/92 where both the Commission and the Bund Naturschutz criticise the Federal German Government over the late and insufficient implementation of the EIA Directive (85/337/EEC).

[32] They highlighted a variety of issues such as acid rain, the construction of motorways and

airport runways, and the dangers associated with nuclear energy. Anti-nuclear protests under the banner of 'Atomkraft nein danke' targeted nuclear plants such as Whyl near Freiburg, Brokdorf near Hamburg, the 'fast breeder' in Kalkar, and Wackersdorf in Bavaria.

[33] The German electoral system combines FPTP and PR: one half of MPs are elected within their constituencies on a direct majority basis, the other half are 'party lists' candidates and are elected via PR. The system allows small and new political parties to enter the parliamentary stage or, at least, threaten the positions of established parties.

[34] In order to avoid unnecessary fragmentation and political disruption, the electoral system includes a '5% hurdle' which small parties have to overcome in order to enter the Bundestag or Länder parliaments.

[35] Environmental considerations were not the only motives behind the German policy on the LCP Directive. Economic level-playing-field considerations (i.e. committing other Member States to the same 'costly' standards) also played a major role in the German position.

[36] C431/92 concerned the extension of a power station in Großkrotzenburg, Hesse; C396/92 concerned the extension of the motorway 'B15 neu' in Bavaria.

[37] Böhmer-Christiansen and Skea (1991) describe the UK Government's anger over the German Government's aggressive strategy during negotiations of the 'Large Combustion Plants' Directive.

[38] In this context, 'energy consumption' serves as an indicator: with over 30 barrels of oil per head in 1991, Germany was just behind the USA (over 50) in terms of energy consumption. Source: UN information in *The Economist* (1996).

[39] In order to enter the third stage of EMU (establishing a European System of Central Banks, a European Central Bank and a single currency), Member States must comply with the following criteria: price stability with an average inflation rate of not more than 1.5% of the 3 best performing Member States over a one year period; no excessive public spending (deficit not more than 3% of GDP); the total government debt should not be more than 60% of the GDP; a stable currency with an exchange rate fluctuating within the margins of +/- 2.25% (within the ERM for at least 2 years); and average nominal long-term interest rates not exceeding more than 2% of the 3 best performing Member States.

[40] English translation: 'Economic location Germany' or 'powerhouse Germany'.

[41] The Brandt Government presented an environmental programme in 1971 which established for the first time a set of environmental principles: the precautionary principle, the polluter pays principle, and the cooperation principle. For more information see Müller (1989).

[42] The Chernobyl accident caused a political crisis in Germany which was later resolved with the establishment of a new ministry for the environment (BMU) in 1986.

[43] The fire at the Sandoz factory near Basle posed pollution and health threats to citizens living in the Southern part of Germany.

[44] For instance, to demonstrate that government measures had improved the water quality of rivers, Töpfer invited the media to watch him swimming in the Rhine.

[45] Data taken from Weidner (1989). Ministers for the Environment do not enjoy a veto right (similar to the Finance Minister's veto right) in policy areas which affect environmental interests.

[46] Under the Grüne Punkt scheme producers are obliged to fund a 'dual system' which collects packaging waste and either recycles the material or disposes it in an

environmentally sensible way.

[47] For more information see, among others, Malunat (1994).

[48] The scope of road traffic predicted in the Federal Traffic Infrastructure Plan of 1985 for the year 2000 had already been exhausted in 1986. For more details see Wissmann (1994).

[49] With over DM 200 billion the Federal Government is funding 12,000 km of new projects and extensions of roads and motorways which, it is hoped, will further generate and strengthen the German economy. Information from Bund Naturschutz (no date).

[50] This research confirms Weale's argument that the national context is a more important influence on environmental politics and policy than are common secular trends (in contrast, Jänicke argues that there is a common, cross-national pattern in the development of environmental policy). See Weale et al (1996).

[51] For discussions on the complex German political-administrative system and its influence on EU policy-making see Rometsch (1995) and Jeffery (1996).

[52] Heritier et al (1994) describe the UK's political-administrative characteristics (such as problems of 'soft regulation' and 'secrecy') and their negative impact on EU environmental policy implementation.

[53] For a detailed account on the divergent standards of EU and German laws, see Lindemann and Delfs (1993).

[54] Another way to describe the UK approach is offered by Heritier et al (1994) who use the term 'konsensuelles bargaining'.

[55] At the time of writing, German legislators are preparing a comprehensive 'Umweltgesetzbuch' which encompasses EU 'integrative' policies such as the IPC Directive. German legislators address these policies with reluctance; they are still considered as 'imported and incompatible policies', 'outcomes of diplomacy which cover up conflicting interests' and 'confused but over-ambitious policy objectives'. See di Fabio (1998). Translation by author.

[56] In fact, Heritier and Knill (1996) notice a shift from German-style EU policies to EU policies which are more compatible with the UK 'integrative' approach.

[57] Ashworth (1992) describes these competencies and emphasises the importance of local government in environmental policy area.

References

Ashworth, G. (1992), *The Role of Local Government in Environmental Protection. First Line Defence*, Longman, Essex.

Blowers, A. (1987), 'Transition or Transformation? - Environmental Policy under Thatcher', *Public Administration*, vol.65, Autumn, pp.277-294.

Böhmer-Christiansen, S. (1990), 'Emerging International Principles of Environmental Protection and their Impact on Britain', *The Environmentalist*, vol.10, No.2, pp.96-112.

Böhmer-Christiansen, S. and Skea, J. (1991), *Acid Politics: Environmental and Energy Policies in Britain and Germany*, Belhaven Press, London.

Bund Naturschutz e.V. (no date), *Der Bund Naturschutz informiert: Verkehrspolitik. Totalschaden!*, Munich.

Commission (1994), *Eleventh Annual Report on Monitoring the Application of Community*

Law, COM (94) 500 final.

Commission (1996), *Thirteenth Annual Report on Monitoring the Application of Community Law*, COM (96) 600 final.

Demmke, C. (1996), *Verfassungsrechtliche und administrative Aspekte der Umsetzung von EG-Umweltpolitik*, European Institute of Public Administration, 30. May.

Di Fabio, U. (1998), 'Integratives Umweltrecht. Bestand, Ziele, Möglichkeiten', *Neue Zeitschrift für Verwaltungsrecht*, Nr.4, pp.329-337.

Gordon, J. (1994), 'Environmental Policy in Britain and Germany: Some Comparisons', *European Environment*, vol.4, part 3, June, pp.9-12.

Heritier, A. et al (1994), *Die Veränderung von Staatlichkeit in Europa. Ein regulativer Wettbewerb: Deutschland, Großbritannien und Frankreich*, Leske + Budrich.

Heritier, A. and Knill (1996), 'Neue Instrumente in der europäischen Umweltpolitik: Strategien für eine effektive Implementation', in Lübbe-Wolff, G. (ed) *Der Vollzug des Europäischen Umweltrechts*, Erich Schmidt Verlag, Berlin.

Jeffery, C. (1996), 'Towards a 'Third Level' in Europe? The German Länder in the European Union', *Political Studies*, vol.44, No.2,, pp.253-266.

Lindemann, H.-H. and Delfs, S. (1993), 'Vollzug des Europäischen Umweltrechts. Lösungsansätze zur Überprüfung und Verbesserung', *Zeitschrift für Umweltrecht*, Nr.6, pp.256-263.

Maloney, W. and Jordan, G. (1995), 'Joining Public Interest Groups: Membership Profiles of Amnesty International and Friends of the Earth', in Lovenduski, J. and Stanyer, J. (eds) *Contemporary Political Studies*, vol.3, Political Studies Association, University of York.

Malunat, B. M. (1994), 'Die Umweltpolitik der BRD', *Aus Politik und Zeitgeschichte* B49/94, 9. December, pp.3-12.

McCormick, J. (1991), *British Politics and the Environment*, Earthscan, London.

McCormick, J. (1993), 'Environmental Politics', in Dunleavy, P. et al (eds) *Development in British Politics 4*, Macmillan, Basingstoke.

Müller, E. (1989), 'Sozial-liberal Umweltpolitik. Von der Karriere eines neuen Politikbereichs', *Aus Politik und Zeitgeschichte*, 47-48/89, 17. November, pp.3-15.

Robinson, M. (1992), *The Greening of British Party Politics*, Manchester University Press, Manchester.

Rometsch, D. (1995), *The Federal Republic of Germany and the European Union. Patterns of Institutional and Administrative Interaction*, University of Birmingham, Discussion Paper in German Studies, December.

Sbragia, A. (1996), 'Environmental Policy: The Push-Pull Policy-Making', in Wallace, H. and Wallace, W. (eds) *Policy-Making in the European Union*, Oxford University Press, Oxford.

Siedentopf, H. (1990), *Die Umsetzung des Gemeinschaftsrechts durch die Verwaltungen der Mitgliedsstaaten*, Europa-Institut, Universität des Saarlandes, Saarbrücken.

Siedentopf, H. and Ziller, J. (eds) (1988), *Making European Policies work. The Implementation of Community Legislation in the Member States*, Sage, London.

The Economist (1996), 'A Survey of Energy', 18 June.

The Economist (1997), 'Road Protests. Victory', 25 January, pp.31/32.

The Economist (1998), 'Environment and transport: Green Gordon', 21. March, p.41.

The Guardian (2001), 'Landfill tax up 10% to deter dumping', 7. March 2001. website (www.guardian.co.uk/)

The Independent (1997), 'Brown puts focus on pollution and energy', 26.November, p.18.

Ward, N., Buller, H. and Lowe, P. (1995), *Implementing European Environmental Policy at the Local Level: The British Experience with Water Quality Directives*, University of Newcastle upon Tyne, March.

Weale, A. et al (1991), *Controlling Pollution in the Round. Change and Choice in Environmental Regulation in Britain and Germany*, Anglo-German Foundation Project.

Weale, A. et al (1996), 'Environmental Administration in six European States: Secular Convergence or National Distinctiveness?', *Public Administration*, vol.74, Summer, pp.255-274.

Weidner, H. (1989), 'Die Umweltpolitik der konservativ-liberalen Regierung. Eine vorläufige Bilanz', *Aus Politik und Zeitgeschichte*, 47-48/89, 17. November, pp.16-28.

Wissmann, M. (1994), 'German Transport Policy after Unification', *Transport*, vol.28A, No.6, pp.453-458.

Young, S. (1993), *The Politics of the Environment*, Baseline Books, Manchester.

5 Environmental Politics and Policy in Scotland and Bavaria

Introduction

The previous Chapter examined EU environmental policies in the context of national (or 'domestic') frameworks. This Chapter refines and magnifies the implementation analysis by focusing on the *sub*national *layer* and its role in the EU environmental policy process. To help understand why *sub*national *layers* process EU environmental policies in a distinct way, the Chapter starts with a general analysis of Scotland and Bavaria's political systems and their positions in the wider national and EU contexts. The Chapter then focuses on Scotland and Bavaria's environmental policies and assesses to what extent their formal and informal *determinants* shape EU environmental policy implementation. It is not argued here that the Scots and Bavarians are necessarily less compliant with EU environmental policy obligations than are UK and German actors in general. Rather, Scottish and Bavarian actors process EU environmental policies in a 'unique' manner and in accordance with their particular political-administrative structures, policy priorities and practices.

The Chapter highlights essentially three reasons why Scottish and Bavarian policy implementation findings differ from their 'mother' states.

- Firstly, actors in Scotland and Bavaria are influenced considerably by their constitutional positions in the wider state systems (i.e. formal *determinant*) which in turn influence formal political-administrative structures as well as informal policy priorities and practices within the regions.
- Secondly, actors in Scotland and Bavaria follow their own informal priorities and attitudes towards environmental protection which are shaped not only by their constitutional embeddedness in the wider national systems but also by *sub*national circumstances such as

Scotland and Bavaria's geographical location, infrastructure, industrial sectors and population density.

• Thirdly, the implementation of EU environmental policies is influenced significantly by the Scots and Bavarians' links with, and attitude towards, the EU and the European integration process.

The Scottish Layer

In order to understand the way in which the Scots have perceived and dealt with environmental matters over the years, it is necessary to establish Scotland's formal position within the UK. Until Scottish devolution in 1999, Scotland had been an integral part of the UK, a centralised state which had developed in a slow and steady evolutionary manner over the past centuries. Scotland's position in the UK was unique: it did not possess its own parliamentary sovereignty and was dependent upon central government which was situated south of the border; yet, Scotland had retained distinctive institutions and enjoyed considerable independence in the areas of education, law and religion.[1] This paradoxical situation changed in the light of Scottish devolution, which introduced a Scottish Parliament and a Scottish Executive and transferred a number of legislative powers to the Scottish level.

Prior to this constitutional reform, Scotland's formal structure and position in the UK political-administrative context was unusual: until 1996, Scottish local government had been divided into Regional and District Councils (in contrast with English and Welsh County Councils), thereafter local government was re-organised into 29 single-tier County Councils. Scottish local authorities co-ordinated (and continue to do so, with the exception of Glasgow City Council) their policies under the umbrella of the Convention of Scottish Local Authorities, in short CoSLA. Apart from co-ordinating Scottish local authorities' views, CoSLA sought to counterbalance the central government ministry responsible for Scottish matters: the Scottish Office.

Formal representation of Scottish interests at large was conducted by the Scottish Office and the Secretary of State for Scotland. Due to the UK's history and constitution, both the Scottish Office and the Secretary of State for Scotland were somewhat 'hybrid' institutions: as parts of central government (i.e. national *layer*), they executed and administered policies which had been legislated in Westminster, while at the same time they represented Scottish interests (i.e. *sub*national *layer* interests) in the UK.

Devolution did not abolish the post of Secretary of State for Scotland. He/she[2] continues to accommodate (and mediate between) UK and Scottish interests.[3] Overall, the dual (reciprocal) representation, while promising close formal ties between the UK and Scottish *layers*, has proven to be a difficult exercise, especially when these interests have been opposed to each other. CoSLA's perceived task to 'counterbalance' Scottish interests by local authorities with Scottish interests by the Scottish Office and the Secretary of State for Scotland confirms that the representation of interests has been rather precarious.

Prior to devolution, the Scottish Office and Secretary of State's 'juggling' of Scottish and UK interests had implications for informal relationships between political-administrative actors North and South of the border. While the Secretary of State for Scotland was (and, in fact, still is) a member of the central government cabinet, he held only the position of a junior minister and therefore enjoyed less political power than his colleagues. The Scottish Office, on the other hand, suffered from a lack of communication with other, sectoral departments. Complaints from civil servants on both sides illustrate this communication problem. Whereas London-based civil servants said about their Scottish counterparts:

We don't know what they do up there – you'll have to ask them,[4]

a Scottish Office official remarked:

We always have to remind London that we exist.[5]

Geographical problems contributed towards this gap: the vast majority of Scottish Office civil servants were located in Scotland while fewer than 1% were based in Whitehall. There were also differences in responsibilities: the Scottish Office pursued territorial interests in an integrated, cross-sectoral manner, while other government departments such as the Department for Transport or the DoE/ DETR pursued sectoral interests. Devolution transformed the Scottish Office into a devolved Scottish Executive, which effectively made these complaints on both sides obsolete (they are no longer part of the same institutional framework). The need to communicate and cooperate, however, still remains, so are the problems in communication between Scottish Executive and Whitehall. The communication gap may even worsen in time as a result of devolved powers, particularly in the area of environmental politics and policy.[6]

Returning to the political scenario prior to devolution, there was another reason for the precarious situation in Scottish representation: until the general elections of May 1997 the majority of Scots did not appreciate their 'Scottish' representation in Westminster. Three-quarters of the Scottish voters had not voted for Conservative Governments and therefore felt that their interests were not represented adequately by Conservative Secretaries of State and a Conservative-led Scottish Office.[7] From 1979 until 1997, Conservative Governments introduced policies such as the council tax and privatisation policies in education, health, energy and housing. These policies met with strong opposition in Scotland.

The discrepancy of informal perceptions dominated relations between the UK and Scottish *layers* and shaped the political-administrative process significantly. Similar tensions between central government and *sub*national regions are evident in every state system. However, in the Scottish case, the dissatisfaction of the Scottish public and many Labour-run (former) Scottish District and Regional Councils towards the UK central government had been more explicit. Apart from electoral discrepancies, relations between the centre and Scottish local governments were tense for two other reasons. In certain policy areas, Scottish local authorities possessed few consultative and discretionary rights and were obliged to receive and execute (unwelcome) Conservative Government policies.[8] In other policy areas such as planning, Scottish local authorities were able to pursue their (economic self-) interests regardless of the UK-wide impact their decisions may have had. This lack of co-ordination on both accounts had a hindering impact on the transposition of policies in general. To a certain extent this problem continues to this day (see below).

As far as links with Brussels are concerned, formal contacts between Scotland and the EU have been dominated by, and were conducted within, the national *layer*. The Secretary of State for Scotland could influence the UK position indirectly through the Cabinet or, in cases that affected Scottish interests in particular, participated in Council of Minister meetings as a UK representative. This practice has not changed with devolution, as relations with the EU remain 'reserved' matters for Westminster and Whitehall. However, the new political structure now includes a First Minister for Scotland and his cabinet colleagues who seek to participate in the process, too. The same rule of indirect, informal influence applies to the former Scottish Office and today's Scottish Executive: Scottish officials have been sent as delegates to the UK Permanent Representation (UKREP) and have been consulted on Commission proposals which concerned Scottish interests (such as the Habitats Directive). In addition, UKREP introduced an information practice which has involved sending short progress reports to the Scottish Office/ Executive. Individual (Scottish) UKREP officials have also maintained regular contacts with the Scottish Office/ Executive.[9] Prior to devolution, co-operation between Scottish and other UK actors on EU matters was marked by suspicion and sometimes mistrust.[10] Following the persistent perception that Scottish interests did not receive adequate attention, Scottish Office officials often felt that their colleagues in other (sectoral) departments ignored Scottish concerns when bargaining in Brussels. Therefore, the Scottish Office maintained links with the Commission by regularly sending 12 to 18 officials to Commission Directorates General (DGs) on secondment. This secondment practice, however, was the result of a voluntary and informal agreement between the Scottish Office and the Commission and did not constitute Scottish representation at the EU level as such. The problem of suspicion and mistrust has been alleviated, although not completely, with more direct Scottish representation in Brussels through the new 'Scotland House' (described below).[11]

Over the years, other Scottish actors such as local government officials have sought independent channels of influence in Brussels and, in contrast to the rest of the UK, have promoted a more 'euro-friendly' image of Scotland.[12] This tradition continues to this day: various Scottish regional, commercial and educational representations lobby EU policy-makers directly and actively.[13] The privately-funded Scotland Europa office, established in 1991, facilitates private and public sector links with Europe. Scotland Europa Ltd represents and informs some 45 subscribing members at the EU level. Apart from lobbyists associated with Scotland Europa, other Scottish actors have sought to influence EU policy-making through European Parliament[14] and through other EU institutions such as the Committee of the Regions. Many Scots working for EU institutions or lobbying for Scottish interest groups have maintained contacts, for instance, via the 'Jock Tamson's Bairns' Index (which lists Scottish officials and employees working in Brussels) or via 'Scotland Yard' (a web page that informs the Brussels community of forthcoming Scottish events).[15] With devolution, another representation was added to the Scottish contingent in Brussels: the Scottish Executive European Office[16] was opened on 'Devolution Day' (1. July 1999) forming, together with Scotland Europa, Scotland House. While Scotland House has been intended as a 'complementary' representation rather than a 'competitive' one,[17] it has nevertheless strengthened the Scottish presence (and confidence) in Brussels.

Despite a number of well-functioning European *feed-back* links (see figure 5.1), Scottish EU representation has been fragmented and has relied (and continues to do so) upon informal contacts. All official EU matters affecting Scotland, for instance complaints about non-compliance with EU policy obligations in Scotland, are still processed via the national *layer*. However, devolution has made a difference in so far as the Scottish Executive can now be held accountable for any cases of non-implementation. In cases of Scottish non-compliance or infringement the Scottish Executive would have to pay fines imposed by the ECJ.[18] In this rather complex scenario, where EU policy influence is limited but implementation and compliance is rather strict, many Scots have started questioning the legitimacy with which the Westminster government continues to dominate proceedings. In the light of the new Labour Government and devolution, tensions may have eased between Scottish and UK interests at the EU level. However, there are indications already that Westminster and Holyrood do not always see eye to eye on political decisions North and South of the border.[19] Moreover, opinion polls conducted since devolution indicate that many Scottish voters re-establish a 'traditional' opposition against Westminster by supporting the Scottish National Party (SNP).[20] At this stage it would be too early to detect any negative impacts of new tensions between Westminster and Holyrood on the *filtering* of EU environmental policies. But the potential exists for problems between the two *layers*; disagreements over EU environmental policies and obstacles hindering their implementation can therefore occur at some stage.

Environmental Politics and Policy in Scotland

Environmental politics in Scotland has been shaped significantly by Scotland's (evolving) formal structure and position within the UK and EU contexts. In particular, Scotland's paradoxical position in the UK (and the EU) and questions surrounding Scottish self-determination and devolution dominated informal perceptions and policy priorities for most part of the 20th Century. Scottish devolution preoccupied political minds in Scotland with the effect that other issues such as environmental problems received not as much attention in Scotland as in other UK (and European) regions. Only in recent years have issues associated with 'green' movements attracted public and media attention in Scotland.[21] The process of adopting and integrating environmental concerns in education, the media and at work has only just started (not entirely as a result of devolution) and many

Scots are increasingly aware of the need to catch up with other Europeans in terms of environmental policy experience.[22]

Scotland's constitutional position has also had an impact on the way (EU) environmental policies have been processed within Scottish political-administrative structures. Prior to devolution, the Scottish Office pursued environmental policy instructions 'from Westminster' while at the same time responding to Scottish priorities and interests. As far as EU environmental policies were concerned, their formal transposition was conducted by the Scottish Office in line with its 'hybrid' position in the UK state system. The Scottish Office generally followed Westminster and adopted 'Scottish' versions of DoE/ DETR Statutory Instruments. While Scottish Statutory Instruments took account of Scottish circumstances (such as the large agricultural and fisheries sectors), they did not differ significantly from their DoE/ DETR counterparts (see in particular case study in Chapter 6). The Scottish Office was traditionally careful not to depart from central government policy and relied upon the guidance from 'down South' on the transposition of EU policies. By and large, the study (and assessment) of EU legislation was neglected by Scottish Office implementors. For instance, when asked about the EIA Directive (85/337/EEC), one Scottish Office official responsible for environmental and planning matters replied that he had 'never read that thing'.[23] This transposition practice was advantageous for EU policies in so far as Scottish implementors did not question or 're-write' EU legislation. On the other hand, DoE/ DETR transposition documents were copied automatically by Scottish implementors without much consideration for the original EU policy texts and without establishing additional measures in support of the policies.

In contrast to the limited Scottish influence in EU environmental policy-making, devolution has changed the way in which Scottish actors formally receive and implement EU environmental policies. It has given the new Scottish Parliament powers to first scrutinise new EU Directives, then consult 'stakeholders' in Scottish society and then formulate its own Statutory Instruments transposing EU policies into the Scottish context. While being formally independent, to date this process has not diverted significantly from that of the rest of the UK.[24] In fact, recently, when it came to the question whether the Scottish Executive should take on a more rigorous regulating role over genetically modified (GM) crops in Scotland, both the Scottish Parliament and Executive signalled that they would not depart from the rest and interfere in this matter; rather Scotland would 'blend in' and leave GM regulation to the UK and EU levels.[25]

As far as internal structures are concerned, prior to devolution the Scottish Office Agriculture, Environment and Fisheries Department was formally responsible for translating (central) environmental policies into the Scottish context. However, while the Department exercised considerable influence in the policy areas of agriculture and fisheries, it showed less determination in pursuing environmental objectives. On the surface, environmental matters were an integral part of the Scottish Office machinery with one Department covering the environment and two other related policy areas. One Scottish Office official argued that he and his colleagues had the advantage of discussing certain policy issues internally and consulting department colleagues on an informal basis. In contrast, the DoE/ DETR had to reach out and approach other departments whenever co-operation and consultation was required.[26] It would therefore appear that environmental matters, and EU integrative environmental policies in particular, were processed more effectively in Scotland than in the rest of the UK. In practice, however, the Scottish Office's approach towards the environment was not exactly 'holistic' as environmental objectives tended to compete (often unsuccessfully) with other interests. On many occasions, the Scottish Office demonstrated a lack of interest in environmental matters, which did not carry the same immediate and lucrative benefits as economic policies.[27] The Scottish Office had the internal means to integrate environmental and other policy matters, but it did not utilise this advantage to provide for a strong policy of sustainable development.

Devolution has not changed this integrated yet half-hearted approach towards environmental objectives. Despite a devolved environmental policy for Scotland and some promising initiatives by consecutive Scottish Executive Ministers for the Environment, the 'historic opportunity' to create a 'holistic' sustainable development policy has not been taken up by Scottish policy-makers. Today, environmental matters are shared between Scottish Executive Departments (in particular the Rural Affairs and Development Departments), a situation that can only hinder the development of a comprehensive and cohesive environmental policy in Scotland. To make matters worse, following the resignation of Environment Minister Sam Galbraith (for health reasons) in March 2001, environmental responsibilities were initially divided and distributed to at least four Executive Ministers (transport, culture and sport, rural development, enterprise). Environmental responsibilities are now shared between two Ministers: Ross Finnie MSP (Minister for the Environment and Rural Matters) and Sarah Boyack MSP (Minister for Transport and Planning). This arrangement still leaves Scotland without a full-time (and some would argue: much needed and deserved)[28] Environment Minister.

Despite the somewhat traditional low-priority stance of environmental issues, by the mid 1990s the Scottish Office had begun to respond to informal pressures from both 'green' NGOs and the European level for more environmental action. While Secretaries of State showed no particular interest in the environment (on no occasion was the environment at the top of their political agenda), environmental concessions suggest that the issue could not be ignored completely. For instance, the postponement of the decision in 1996 to build a second Forth road bridge by Ian Lang (Secretary of State for Scotland until May 1997) suggested that the Scottish Office did adapt to 'green' pressures. Scottish Office Ministers for the Environment were not enthusiastic towards the 'green cause'. In particular, Sir Hector Monro refused to open traditional lobbying networks to include environmental NGOs in Scotland. His successors, the Earl of Lindsay and Lord Sewel, signalled a change in attitude by showing an interest in environmental issues and initiatives.[29]

Since devolution, policy documents such as *The Nature of Scotland* and Sam Galbraith's decision to refuse planning permission for the controversial Harris 'superquarry'[30] have given the impression that the Scottish Executive is more 'environmentally friendly'. However, other, documents such as the Executive's document on climate change (which disappointed many 'green' activists) and the decision not to fill the post of Environment Minister have indicated that First Minister Henry McLeish and his colleagues have taken a lukewarm interest in 'green' issues. Overall, however, consecutive Scottish Office and Scottish Executive Ministers have felt obliged to respond to some of the pressures coming from Scottish environmentalists and Brussels. It can therefore be deduced that some (slight) policy changes do take place in Victoria Quay.[31]

In order to accomplish environmental policies successfully, formal and informal links between the Scottish Office/ Executive and Scottish local authorities should ideally be close and co-operative. However, as in most other areas, environmental policy links between the government levels have shown weaknesses pre- *and* post- devolution. Prior to devolution, links were dominated by an overall mistrust between the (Conservative-led) Scottish Office and (predominantly Labour-led) local Councils and could be described as non-cooperative. Relations were not helped by measures such as the 1980s' privatisation policy introduced by the Conservative Government, which deprived Scottish local authorities from an effective environmental policy tool by privatising bus services in Scotland. According to a (former) Regional Council official, local authorities effectively lost the control over public transport, which could have been used to establish a more environmentally-friendly, sustainable infrastructure.[32]

Since devolution, relations have evolved: tensions are not so much dominated by party politics but more by the Scottish local authorities' wearyness to what extent Scottish Parliament and Executive can take political powers (and financial resources) away from their own level. There have been improvements, however. For instance, shortly after devolution, CoSLA and the Scottish Executive established a new 'partnership' intended to strengthen links between the two levels (see below). On the other hand, Council officials have already expressed their disappointment in the Scottish Executive's lack of environmental policy leadership.[33] It remains to be seen to what extent the new 'partnership' will further shape environmental politics and policy in Scotland.

Apart from the informal factors that determine the relationship between government levels, there are other, formal, *determinants* that influence the practical application of EU environmental policies. As mentioned above, Scottish local authorities have enjoyed considerable discretionary room over certain policy areas, including the practicalities of many EU environmental policies (such as determining EIA projects, see Chapter 6). Yet, the Secretary of State for Scotland and since devolution the First Minister (or one of his colleagues, depending on environmental policy issue) have retained their ultimate right to over-rule local authorities' decisions. While this 'over-ruling' authority has remained, discretionary powers in other areas have meant that implementation performances have been less co-ordinated and have depended on priorities and relationships between actors inside local authorities. As a result, the implementation of EU environmental policies has depended upon individual administrators and their interpretation of EU legislation and Statutory Instruments, unless, of course, the Secretary of State/ First Minister interfered with their decisions.

In this constellation, CoSLA has tried to provide an effective link of co-operation between the Scottish Office/ Executive and Scottish local authorities in the environmental policy area. For example, in 1993 Scottish Office and CoSLA representatives published a *Local Environment Charter for Scotland,* which, however, met with harsh criticism from environmental officers within the Councils. Many Council officials were outraged over the perceived arrogance with which the Charter was put forward and criticised the fact that only a small number of Council officials were involved in the Charter's preparation. Reactions such as these confirm that the working relationship between the Scottish Office and Scottish local authorities (despite CoSLA's efforts) was marked by mistrust. An improvement in relations became noticeable, however, after the Labour victory of May 1997, when Labour-run Scottish local authorities started supporting CoSLA in its 'sustainable Scotland' partnership with the (then) Scottish Office.[34] This new 'partnership' (mentioned above) is still in the process of evolving but in time could prove beneficial for the implementation of EU environmental objectives as actors at both Scottish and local community levels express their interest in closer co-operation.

Focusing on Scottish local authorities, environmental politics and policies have shown strengths and weaknesses. Environmental objectives have suffered under formal constraints: the 1996 local government re-organisation in Scotland involved a major administrative shake-up from a two-tier to a single-tier local government system and occupied local administrators for several months. The shake-up and redundancies caused by the merging of responsibilities of former Regional and District Councils pushed many policies to the side-line of local politics.[35] Similarly, the implementation of EU environmental policies did not receive top-priority attention during the process.

Prior to re-organisation, the majority of the District and Regional Councils produced environmental programmes (or charters), which outlined the Councils' commitment in protecting the environment.[36] These documents usually listed internal measures, for example energy saving and recycling schemes, and measures which promoted environmental awareness outside the Councils.[37] While environmental initiatives and contacts with environmental organisations were disrupted by the local government shake-up, many informal initiatives and contacts were taken up again by environmental officers within the Councils. Over the years, environmental officers have consulted NGOs such as Friends of the Earth Scotland (FoEScotland) and have maintained close links with 'green' groups in Scotland. However, environmental officers have often felt isolated and have complained about their colleagues' lack of interest in 'green' issues.[38] Today, environmental officers enjoy considerable freedom in producing environmental charters and programmes; they also participate in Local Agenda 21, which follows Rio's Agenda 21 to promote sustainable development at the local community level.[39] However, as far as influencing other local government departments is concerned, environmental officers continue to play a minor role in local authority activities; they are 'tolerated' but not integrated as influential actors in Scottish local politics.[40]

One environmental officer argued that Scottish local authorities have followed a low key environmental policy approach because public opinion in Scotland has not put enough 'green' pressure on them.[41] This lack of 'real pressure' might also explain the Scottish Executive's lukewarm attitude towards environmental issues. In the past, the public's preoccupation with Scottish devolution helped explain the somewhat half-hearted interest in 'green' issues. However, devolution cannot explain why 'green' issues continue to be 'low key'.[42] Another possible explanation points towards Scotland's rich natural resources, which have been taken for granted by many Scots. Because of Scotland's relatively low population density and an abundance of natural resources, pollution problems have been less visible and have therefore appeared to be less urgent. At the same time, Scotland's natural resources have supported large economic sectors such as the wool and whisky industries and tourism. For this reason, many Scots have been 'unconsciously aware'[43] of environmental issues. They may have shown an interest in natural resources (such as water), but mainly because they constituted essential components for their economy.

A 1995 public opinion survey concluded that while the majority of respondents in Scotland were generally concerned about environmental problems, they quite strongly favoured economic interests over other considerations.[44] Many respondents were not prepared to restrict economic development such as the building of out-of-town shopping centres. 73% of respondents agreed that:

> if people want to go shopping in their car, it's up to them.

The Brent Spar controversy of 1995 also suggests that the majority of Scots have been less worried about environmental problems than other Europeans. The planned disposal at sea of the 'Brent Spar' Shell oil rig caused more public opposition in other European regions than in the region most directly affected by the plan: Scotland and particularly the north west of the Hebrides.[45] One of the key reasons behind the Scots' reluctance to challenge a large company (and employer) can be found in Scotland's economic conversion from old heavy industries to new industries such as North Sea oil exploitation. Enjoying material wealth, many Scots have been reluctant to restrict immediate economic development for the sake of less tangible environmental benefits.

Scots have also perceived Scotland's peripheral situation in the Single Market as a disadvantage that has to be compensated with lenient environmental standards and the promotion of economic development.[46]

Indeed, many Scottish actors from the public and private sectors have opposed EU environmental policies, which seemed to threaten economic opportunities in disadvantaged regions and imposed financial and administrative costs, costs which are perceived as disproportionally high for Scotland.[47] In fact, many officials at both Scottish Office/ Executive and local authority levels have abandoned their 'euro-friendly' attitude when faced with 'expensive' EU environmental policy obligations. For instance, the Drinking Water Directive and the 'voc Stage I' policy of 1984 (which regulates the capturing of emissions at petrol stations) met with strong opposition in Scotland. Scottish Office officials and those affected by the policies (such as farmers with their own water supplies and petrol station owners) felt that the periphery situation of the Highlands constituted an unacceptable disadvantage as remote areas could not afford the changes to drinking water supplies and petrol stations.[48] Pragmatic economic (self-) interests have therefore influenced key actors' perceptions towards EU environmental policies. Faced with 'inconvenient' EU obligations, many Scots have tended to complain as much about the costs imposed by 'bright-eyed, bushy-tailed junior Commission officials' as their colleagues in other EU regions.[49] Therefore, the above described close informal relations between EU and Scottish actors dampen as soon as policy details over potential economic restrictions and sacrifices come to light. Unless they fit conveniently into existing procedures and priorities, EU environmental policies face a lack of commitment, if not down-right resistance, from many Scottish implementors. Comments such as the following illustrate this concern over EU-imposed burdens and come as no surprise.[50]

The most endangered species in the Highlands is man.

However, it would be misleading to assume that the Scots have only been interested in economic benefits. A number of Scottish institutions and organisations have fulfilled key monitoring, educational, informative and advisory functions in the environmental policy area. Since 1996, the quango Scottish Environmental Protection Agency (SEPA) has pursued the tasks of pollution control and waste regulation.[51] SEPA has been criticised for being an over-centralised institution, for not ensuring the representation of local authorities and for neglecting vital monitoring tasks such as ICM.[52] SEPA officials have responded to the criticism by stressing that the agency is still in a process of learning and identifying pollution control criteria. Some SEPA officials have continued with the former HMIPI and RPBs' neutral and passive approach while many others have taken a more 'green-activist' position (at least at public meetings) in Scottish environmental politics.[53] In terms of EU environmental policy implementation, however, SEPA has played an important monitoring and control function in the areas of air quality control as well as waste and water management.[54]

Scottish Natural Heritage (SNH) has fulfilled advisory and monitoring functions in Scotland and has entertained close ties with the Scottish Office. SNH has been criticised by environmental activists, however, for acting like a central government agent and adopting (the former) Conservative Government terminology such as 'efficiency' and 'value for money'.[55] Nevertheless, SNH has contributed towards environmental awareness in Scotland and has taken the lead in a number of environmental initiatives.[56] With regard to EU environmental policies, SNH has contributed towards the implementation of the 1979 Wild Birds Directive and, more recently, has played a leading role in the first implementation stage of the EU Habitats-Directive (92/43/EEC).[57]

The environmental NGO Friends of the Earth Scotland (FoEScotland) has regularly attracted public and media attention with campaigns such as 'Slow down Scotland' in 1994 and has maintained contacts with the EU Commission on a number of environmental issues.[58] FoEScotland have utilised their links with the Commission as 'headline grabbers' and have thereby strengthened the 'green cause' in Scotland.[59] FoEScotland and other NGOs have welcomed the environmental input from the EU *layer* and have put pressure on Scottish implementors to comply with EU environmental obligations.

Finally, there are a number of organisations in Scotland which have fulfilled informative functions. For instance, until its demise due to lack of (government) funding, the Scottish Environmental Education Council (SEEC) co-ordinated the work of 'REEFs' and informed children in Scotland about environmental issues by raising 'green awareness' in Scottish schools.[60] Similarly, the Centre for Environment and Business in Scotland (CEBIS), funded by Scottish Enterprise and membership fees, informed interested businesses of the latest developments in UK and EU environmental legislation which affect Scottish industry.[61] In doing so, CEBIS made a valuable contribution towards the implementation of (and compliance with) EU environmental policies.

Figure 5.1 summarises the Scottish *layer* and its formal and informal *determinants* which have shaped (and continue to shape) (EU) environmental policies in Scotland. Scotland has featured some favourable *determinants* (such as economic sectors depending on 'healthy' natural resources) as well as unfavourable *determinants* (such as tensions between key actors and a 'half-hearted' approach towards pollution problems) for the environmental policy process. Recent years have seen a (very) moderate shift towards environmental awareness in Scotland. Representatives from a wide spectrum of institutions and interest groups now seek to strengthen policy links between actors and government levels and establish a 'Scottish' environmental policy.[62] Yet, despite the 'historic opportunity', provided by Scottish devolution, to establish a coherent (and effective) environmental policy, latest developments have shown that other policy priorities (such as pensions and jobs) continue to receive more attention in Scotland than do 'green' issues such as recycling and public transport.[63]

In terms of EU environmental policy implementation, some Scottish actors (such as 'green' NGO activists) have welcomed and actively supported EU environmental policies while many others from the public and private sectors have openly expressed concern over the additional costs and work associated with EU environmental obligations. These 'burdens' have been perceived as particularly unfair for Scotland, a region which is at the periphery of the Single Market. While EU environmental policies have benefited from a comparatively swift and uncomplicated transposition process in pre-devolution Scotland, they have faced resistance from conservative (economic) priorities in Scotland. These priorities have often been incompatible with EU environmental obligations and have presented obstacles in the implementation path. The formal structures and with them the formal processing of EU environmental policies have changed with devolution, however, informal relationships and attitudes towards environmental issues have not changed significantly. As a result, devolution has only had a limited impact on the actual *filtering* of EU environmental policies and their outcomes in Scotland.

Figure 5.1: The Scottish Layer and Environmental Policy

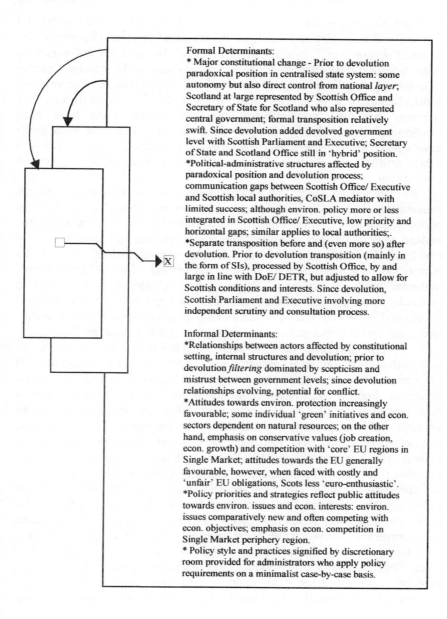

Formal Determinants:
* Major constitutional change - Prior to devolution paradoxical position in centralised state system: some autonomy but also direct control from national *layer*; Scotland at large represented by Scottish Office and Secretary of State for Scotland who also represented central government; formal transposition relatively swift. Since devolution added devolved government level with Scottish Parliament and Executive; Secretary of State and Scotland Office still in 'hybrid' position.
*Political-administrative structures affected by paradoxical position and devolution process; communication gaps between Scottish Office/ Executive and Scottish local authorities, CoSLA mediator with limited success; although environ. policy more or less integrated in Scottish Office/ Executive, low priority and horizontal gaps; similar applies to local authorities;.
*Separate transposition before and (even more so) after devolution. Prior to devolution transposition (mainly in the form of SIs), processed by Scottish Office, by and large in line with DoE/ DETR, but adjusted to allow for Scottish conditions and interests. Since devolution, Scottish Parliament and Executive involving more independent scrutiny and consultation process.

Informal Determinants:
*Relationships between actors affected by constitutional setting, internal structures and devolution; prior to devolution *filtering* dominated by scepticism and mistrust between government levels; since devolution relationships evolving, potential for conflict.
*Attitudes towards environ. protection increasingly favourable; some individual 'green' initiatives and econ. sectors dependent on natural resources; on the other hand, emphasis on conservative values (job creation, econ. growth) and competition with 'core' EU regions in Single Market; attitudes towards the EU generally favourable, however, when faced with costly and 'unfair' EU obligations, Scots less 'euro-enthusiastic'.
*Policy priorities and strategies reflect public attitudes towards environ. issues and econ. interests: environ. issues comparatively new and often competing with econ. objectives; emphasis on econ. competition in Single Market periphery region.
* Policy style and practices signified by discretionary room provided for administrators who apply policy requirements on a minimalist case-by-case basis.

The Bavarian Layer

In order to understand how Bavaria has processed EU environmental policies, it is necessary to investigate the Bavarian *layer* and its formal and informal *determinants*. In contrast to Scotland, Bavaria has not witnessed major constitutional changes in its formal position in the Federal German state system; the 'Free State' of Bavaria[64] has shared with the Federal level political-administrative competencies since 1949. The federal set-up of Germany has had two effects: both government levels have been 'verflechted' (intertwined) with each other, at the same time both levels have tended to compete with each other over responsibilities.[65] Over the years, attitudes in Bavaria have very much confirmed this notion of competition.

In theory, once the Federal level (i.e. the Bundestag and the Bundesrat) adopts a Federal Law, the Länder are obliged to execute the policy according to the principle of 'federal loyalty' (Art.83). In return, the Federal level is obliged to respect the legislative and executive powers of the Länder. In practice, this careful balancing of Federal and Länder powers has been a cumbersome process and has placed formal and informal obstacles in the policy path. Bavarian political-administrative actors in particular, have insisted on their autonomy in many policy areas and have maintained a strong and well-functioning political-administrative system which has resisted 'instructions from above' as much as possible.[66]

To illustrate the Bavarian determination to establish an independent political level, Bavarians adopted their own constitution in December 1946, one year after the Second World War and three years before the Federal constitution came into being. The Bavarian constitution stipulates the principles of democracy and self-determination as well as the Bavarian national heritage and Christian (predominantly Catholic) values. Policy-making in Bavaria has been an elaborate democratic process of several readings, involving the Bavarian State government, a lower chamber (i.e. the Land parliament or 'Landtag'), and an upper chamber (i.e. the Bavarian 'Senat').[67] In some specific cases, which involve substantial reforms, the Bavarian policy-making system has even allowed for referenda, a measure which is unusual for German politics.[68] The Bavarian State government consists of a 'Ministerpräsident' and nine government ministers. The Ministerpräsident, who is supported by the 'state chancellery' of about 330 officials, has traditionally enjoyed a powerful position, with the other nine ministers and their sectoral departments[69] following his[70] political lead. Despite the two chamber system, Bavarian

Ministerpräsidents have tended to be charismatic and powerful politicians who have enjoyed the full support of their political party, the CSU. The CSU has been, for German standards, an unusual political party; it represents only territorial (i.e. Bavarian) interests and secures vast majorities at every Bavarian election.[71]

Bavarian State governments have tended to compete with Federal Government and have not shied away from occasional confrontations with the Federal level and other Länder.[72] Although constitutional obligations have been followed with 'federal loyalty', Bavarian politicians have tended to perceive 'outside' obligations as disturbances to Bavarian affairs. The CSU Government in particular, has promoted the idea of Bavaria's self-sufficiency and has been reluctant to commit Bavaria's resources to projects outside its range of powers. Co-operation with other European regions has been generally welcome, however policy 'instructions' from the national and EU *layers* have been perceived as unnecessary 'burdens' which should be avoided.[73]

Along similar lines, local authorities in Bavaria have enjoyed discretionary powers based on the Bavarian constitution. The Bavarian 'Free State' is organised in a three-tier government system with the Bavarian State government level at the top of the ladder, the seven Districts (Regierungsbezirke)[74] at the intermediate level, and the County (71 Landkreise) and Town (25 Kreisfreie Städte) Councils at the bottom of the system.[75] In order to extract a common position which represents policy interests and opinions, each local government level has organised itself within umbrella organisations (Spitzenverbände)[76] equivalent to the Scottish CoSLA.

In contrast to CoSLA, however, Bavarian Spitzenverbände have been in the formal position to ensure that no government level 'above' oversteps political-administrative boundaries. Occasionally, Bavarian local authorities have had to protect their 'autonomy' when confronted with a dominant Bavarian State government.[77] Consequently, informal relations between Bavarian local authorities and the Bavarian State government have not been entirely co-operative. In fact, a former Bavarian local authority spokesperson described relations between the Bavarian State government and local authorities as 'stiefmütterlich'.[78] This relationship has been worsened by economic and financial pressures following German unification, the 1990s' recession and economic targets associated with the EMU.[79] The Bavarian State government has responded to these pressures by shifting a number of 'expensive' responsibilities to lower government levels.[80] Bavarian local authorities in turn have had to cope with public criticism over unpopular and stringent economic decisions. As a result, communication between the Bavarian State government and local authorities has deteriorated further.[81]

Apart from disputes between the Bavarian State government and local authorities over competencies, both levels have pursued different objectives. The Bavarian State government has focused on the formulation of legal texts for Bavaria, while Bavarian local authorities have been occupied with the subsequent practicalities of policy obligations. While this sharing of tasks makes sense, the difference in perceived objectives has constituted a psychological gap between 'instructors' and 'implementors' which does not facilitate the policy process in Bavaria.

Tensions between government levels have also been evident in Bavaria's formal and informal links with the EU. The Bavarian State government has been one of the key campaigners for increased *sub*national participation at the EU level. Bavaria hosted a number of Länder and European regions conferences[82] which contributed towards the adoption of policies such as the principle of subsidiarity and the establishment of the Committee of the Regions.[83] Following Länder pressure, the Federal Government had to accept constitutional adjustments in the form of a new Article 23[84] which has allowed Länder formal representation at Council meetings whenever Länder competencies are affected.[85] While Article 23 has strengthened the formal position of the Länder in the EU process, a clearer line still has to be drawn establishing exactly which government level participates at which Council meeting. In addition, Länder governments have already signalled their continuing dissatisfaction with the inadequate participatory powers at the EU level.[86] It remains to be seen whether Bavaria and the other Länder can utilise their new powers and shape EU policy-making effectively. So far, the 'fusion' of Federal and Länder representation at the EU level has resulted in 'confusion' over competencies and interests and could well lead to a slow-down of the EU environmental policy process.[87]

Both the Bavarian State government and Bavarian local authorities have maintained well-resourced quasi-representation offices in Brussels: the 'Informationsbüro des Freistaates Bayern' and the 'Europabüro der Bayerischen Kommunen'.[88] The 'Freistaat' office lobbies EU institutions on behalf of the Bavarian State government and focuses on economic interests and formal implementation problems, while the 'Kommunen' office represents Bavarian local authorities and their every-day practical problems with EU policies. Apart from the two Bavarian offices in Brussels, two members (out of 24 German members) of the

Committee of the Regions (CoR) represent Bavarian State interests, while one CoR delegate (out of three 'Kommunen' delegates) represents Bavarian 'Kommunen' interests and fourteen MEPs come from Bavarian constituencies.[89] Bavarian lobbying techniques have been criticised by officials from EU institutions and Federal German representations as 'awkward' and sometimes even 'intimidating'.[90] Bavarians have enjoyed comparatively strong formal links with the EU (see *feed-back* in figure 5.2) and have had the (financial) resources to be heard at the EU level, but their 'awkward' informal lobbying has often hindered effective Bavarian participation in EU politics.

Environmental Politics and Policy in Bavaria

> In Bayern gehen die Uhren anders
> In Bavaria the clocks run differently

This often-quoted phrase not only describes Bavaria's unique position and internal structures in general, the phrase also applies to the unique features of Bavaria's environmental politics and policy. In contrast to Scotland, Bavarians have been interested in 'green' issues from an early stage and have used their formal means to establish an environmental policy which has differed in many respects from the Federal German policy. In fact, the Bavarian 'Free State' included environmental protection as one of its state objectives[91] long before the Federal level decided to amend the German constitution.[92] Apart from the 'green' amendment of Bavaria's constitution, Bavaria established an additional principle of 'environmental precedence': environmental concerns shall take precedence over other considerations such as planning.[93] Other Bavarian forerunner initiatives include –

- the first school curriculum which mentions environmental education (1976)
- the first systematic assessment of natural habitats (called 'Biokartierung') from the mid-1970s onwards
- the first comprehensive and systematic measuring networks in the areas of air quality, soil and nuclear energy.[94]

Since Bavarians have been able to establish an independent environmental policy at an early stage, it should be an easy task to identify a 'Bavarian' environmental policy and a 'Bavarian way' of EU environmental policy implementation.

At the Bavarian State level, the Bayerische Staatsministerium für Landesentwicklung und Umweltfragen (StMLU),[95] established in 1970 and the first ministry of its kind, has played a central if not dominant role in Bavaria's environmental policy. Untypical for the German (and Bavarian) compartmentalised policy approach, the StMLU was established to accommodate two vital interests: economic development on the one hand and rural-traditional values on the other hand.[96] To a certain extent, this underlying concept continues to apply to this day as StMLU officials still consider their role as balancing and combining environmental protection with economic development. In this sense, the StMLU has been the first government department in Europe to apply the approach of environmental policy 'integration'. However, in terms of EU environmental policy implementation, the StMLU has not been a 'truly integrative' institution. In particular, the StMLU has had serious misgivings with broad and cross-sectoral policies such as the EIA Directive or the IPC Directive. These integrative policies have been processed in a fragmented, piecemeal and even technocratic manner. Communication and co-ordination between the StMLU and other sectoral departments and government levels over the implementation and effectiveness of these and other EU environmental policies have been almost non-existent.[97] A coherent picture (and further strengthening) of EU environmental policies is therefore difficult to achieve.

Over the years, the StMLU managed to give equal weight to environmental and economic aspects. At least, citizens and 'green' NGOs in Bavaria appeared to be satisfied with the handling of environmental matters. However, since the early 1990s, the economic-environmental balance has drifted towards economic priorities and the StMLU has been criticised by 'green' NGOs for compromising the environmental policies of the 1970s and 1980s in favour of economic deregulation of the 1990s.[98] The Bavarian Development Programme of 1993, for instance, deleted key clauses such as the following:

> A healthy environment should not be sacrificed for the sake of economic growth.

In addition, statements such as 'policy objectives are to be implemented' were replaced by vague formulations such as 'should be pursued if possible'.[99] The high-profile *Umweltpakt* of 1995 also signalled a major policy change.[100] Following a new 'substitution and deregulation' approach,[101] the StMLU 'eased the financial and administrative burden' for businesses caused by environmental control. In return, private sector representatives committed themselves towards a number of 'voluntary' environmental obligations listed in the pact document. Having enjoyed considerable popularity with businesses in the first five years, the StMLU launched another follow-up *Pakt* in October 2000, which was signed by some 900 business representatives from a broad range of sectors including tourism, manufacturing and the service sector.[102] While business representatives welcomed the pact with open arms, 'green' NGOs such as the Bund Naturschutz (see below) expressed their concerns over the resulting deregulation below acceptable (legal) standards.[103]

While the StMLU has conducted a strategic change from a stringent to a more flexible and lenient environmental policy, it continues to supervise and co-ordinate a variety of environmental activities in Bavaria. Its role in Bavaria's environmental policy is so central that policies from the EU and national *layers* are received with considerable reluctance. In fact, the StMLU has made no secret of its opposition against policies from 'outside' which it considers unacceptable and incompatible with Bavarian standards.[104] This rather defensive behaviour of the StMLU has represented an obstacle in the path of environmental policies which derive from the national and EU *layers*. The complex sharing of powers between the Federal and Bavarian levels has contributed towards the perception that every EU environmental policy has to undergo a scrutiny process. In the process, EU environmental policies are shaped to fit Bavaria's existing legal-administrative system.

Following the above described tradition to confront other government levels, the StMLU has tried to resist many 'inconvenient' EU environmental policy obligations. StMLU officials have tended to wait until Federal and Länder colleagues completed their EU policy tasks and have frequently blamed the Federal level for their own transposition delays.[105] Arguing that the implementation of most EU policies lie within their exclusive domain, Bavarian implementors have also considered it unnecessary to inform Federal and EU actors on their transposition performances and have insisted that both EU and Federal levels are not in the position to supervise Bavaria's compliance with EU obligations.[106] This attitude was confirmed recently when StMLU Minister Schnappauf denied non-compliance with the Wild Birds Directive and instead attacked the Federal Minister for the Environment, Trittin. According to Schnappauf, Trittin was not in the position to 'remind' Bavaria of its legal obligations. In fact, the Federal level was not in a controlling position but was 'merely a messenger' of Länder information to Brussels. Pointing the finger at Bavaria was an 'insult' considering that the Federal level was behind schedule with its EU legal obligations.[107]

The Bavarian State government has been adamant to maintain this unwritten rule of implementation autonomy and has opposed any form of instruction concerning the transposition and practical implementation of EU policies. Bavarians have also questioned the necessity of many EU policies on the grounds of subsidiarity.[108] This 'euro-sceptical' attitude, an attitude which is unusual for Germany as a whole,[109] is unlikely to change in the near future and will continue to influence relations between Bavarian and EU actors during environmental policy-making and implementation.[110]

Despite the StMLU's central co-ordinating position, links between the StMLU and government levels below have suffered from the above described tensions over competencies. The Bavarian State government has responded to the economic pressures of the 1990s and has shifted the burden of financing many environmental policies to Bavarian local authorities, which in turn have had to deal with expensive pollution control measures in areas such as noise, air, soil, waste and water.[111] Not surprisingly, Bavarian local authorities have considered these responsibilities and further environmental instructions from 'above' not so much a necessity rather a punishment which causes considerable financial and administrative problems.

The Bavarian Spitzenverbände have taken a passive stance in the co-ordination of environmental policies. One of the few inter-regional initiatives was taken in May 1997 when a private marketing company was commissioned by the Bavarian Landkreistag to survey local authorities' environmental activities and prepare a comprehensive overview of environmental policies in Bavaria.[112] Under the guidance of the Landesamt für Umweltschutz (LfU), local authorities have also participated in a Local Agenda 21 network called 'KommA21' which provides a platform for information exchange.[113] While these and other initiatives have been intended to strengthen policy co-operation and co-ordination between the regions (and government levels), local authority officials still prefer to compartmentalise environmental matters along horizontal and vertical lines. For instance, the pollution control of a river is divided into sections according to size and relevance for infrastructure, with different government levels taking care of different parts of the river. This sectoralisation and fragmentation has implications for the pursuance of many environmental policies: efforts to comply with environmental objectives are not co-ordinated between government levels and departments and there is a general lack of transparency over policy results. Ultimately, inadequate information and lack of transparency contribute towards implementors' lack of commitment in environmental policies and particularly policies which require a joint effort and affect several policy areas.

The fragmentation and lack of a coherent overview is reflected in local authorities' environmental programmes and charters. Few policy documents are available in Bavaria and those that are available, are not comprehensive but instead focus on either nature protection[114] or the technological side of pollution prevention and control.[115] Apart from these specialised areas, there are also a number of leaflets in circulation which outline individual local community initiatives for the environment.[116] In many instances, local authority officials referred the author to StMLU publications such as *Die Umweltbewusste Gemeinde* (1996a), an information pack (compiled with the help of Bavarian local authorities) which provides information on existing environmental initiatives and gives advise on setting-up new initiatives. To date, Bavarian local authorities have not produced a common environmental strategy which would help co-ordinate and strengthen cross-boundary and cross-sectoral environmental policies.

Considering Bavaria's overall environmental policy, not only formal structures have influenced environmental institutions and policies, other *determinants* such as informal attitudes towards environmental issues, traditional values, economic priorities and geographic location have shaped institutions and policies. Environmental awareness has been particularly influenced by Bavaria's traditional values which constitute an essential part of Bavarian politics. Bavaria has often been described as a 'Flächenstaat' - a state with large rural and agricultural resources as well as forests and parks.[117] In addition, traditional economic sectors such as farming, tourism and beer brewing have relied upon a healthy environment. There are obvious similarities with Scotland. However, in comparison with Scotland, Bavaria's attitude towards natural resources and its 'conservatism' have combined successfully rural traditions and nature-orientated values with more recent 'green' issues such as problems of air pollution.

This combination of traditional and new environmental concerns contributed towards the early establishment the Bund Naturschutz in Bayern e.V.[118] in 1913. Over the years, the Bund has fulfilled informative and advisory functions similar to the Scottish SNH. However, in contrast to SNH, the Bund has been financially independent from government and has played a more confrontational role. With its campaigns such as the promotion of small-scale farming methods, which allow for biodiversity and the preservation of rural communities, the Bund has merged Bavarian traditional-rural values with new 'green' issues.[119] The Bund has also campaigned for the protection of the Alpine and Danube regions which represent valuable parts of the Bavarian heritage.[120] The Bund has collaborated with other environmental NGOs such as the Landesbund für Vogelschutz in Bayern e.V. (LBV).[121] The LBV, too, has promoted the protection of Bavaria's natural heritage, in particular the protection of Bavaria's wild birds such as the kingfisher.

The Bund and the LBV have co-operated with the EU Commission on a number of occasions and have used EU institutions and policies to support their 'green' campaigns.[122] Both NGOs have sought to influence the implementation of EU environmental policies such as the Habitats Directive and sometimes have had to initiate legal proceedings against the Bavarian State government. For instance, the Bund was involved in a legal case against the Bavarian State government which concerned the question whether the 'B15 neu' motorway project was compliant with the EIA Directive. LBV complained bitterly about the last-minute consultation by the StMLU on the list of protected areas for the Habitats Directive.[123] Bavarian environmental NGOs possess comparatively large membership and financial resources, but their quasi-exclusion by the Bavarian State government from the implementation process can only be disadvantageous for EU environmental policies.

Another key actor in Bavarian environmental politics is the Bayerische Landesamt für Umweltschutz (LfU). The LfU exemplifies the Bavarians' preference for measuring and assessing pollutants and environmental impacts in Bavaria (see informal *determinant* in figure 5.2). The measuring of qualitative and quantitative environmental standards has been elaborate and extensive in Bavaria. Particularly the LfU, Bavaria's equivalent to SEPA, has monitored environmental standards in great detail and has provided information for the public and private sectors as well as interested citizens. The LfU has collected data on air, soil, water, noise, waste, nature, and nuclear safety, and has prepared assessments and environmental reports such as the 'Biokartierung'. It was the first agency to set up a centralised air quality monitoring network in 1974 and now supervises 73 monitoring stations. In addition, the LfU today runs 30 stations specifically measuring radioactivity in air. With the new LfU Headquarters in Augsburg (set up in 1999 and employing 430 officials), the Bavarian State government intends to create a centre of 'green technology', which provides information and know-how particularly for the industrial sector.[124]

The attribute of technological perfectionism has been quite pronounced in Bavaria, perhaps even more pronounced in Bavaria than in the rest of Germany (and, indeed, the rest of the EU). Bavaria has been one of the wealthiest regions in the EU; its citizens have been able to accept and afford stringent technological standards.[125] On the other hand, the focus on 'water-tight' technological standards has also meant that many EU environmental policies that are less measurable (i.e. policies such as the EIA policy) fell in the 'incompatible' category and were neglected by Bavarian implementors because they did not fit into Bavarian standards and practices.

Since the early 1990s the Bavarians' vigourousness with environmental standards has taken a turn with increasing economic pressures. German unification together with the opening towards Central and Eastern Europe, ongoing economic pressures and tough EMU targets have changed informal policy priorities in Bavaria. While other EU regions and German Länder have faced the same or similar economic pressures, Bavarians have perceived these problems as particularly burdensome and the most difficult problems since the Second World War. The prospect of economic instability, partly caused by economic and political changes in neighbouring countries to the East, have compelled the Bavarian public and politicians to consider substantial economic policies. The Bavarian State government has responded to economic fears by launching initiatives such as the 'Bayerische Offensive' (which invests public money in new businesses) and the above mentioned *Umweltpakt*.[126] The Bavarian State government now relies heavily upon the private sector's voluntary self-discipline (in German: Eigenverantwortung) to limit environmental pollution and develop clean technologies.

With the economic pressures of the 1990s, Bavarian political-administrative actors and representatives from the private sectors have distinguished more carefully between 'welcome' and 'unwelcome' EU environmental policies. 'Inconvenient' policies involving administrative changes and adjustments of threshold criteria have caused headaches for Bavarian administrators, while policies based on voluntary action such as the eco-audit Directive have been looked upon favourably in the light of Bavaria's deregulation measures. Nevertheless, the majority of EU environmental policies are still perceived as unnecessary, incompatible with Bavarian standards, and 'imposed by outsiders' who do not have a legitimate right to do so.[127] Comments such as the following are typical for the Bavarian State officials' attitude towards EU institutions and EU environmental policies:

> Brussels does not even have a proper sewage plant, so who are they to set high standards for us?![128]

Bavaria has been at the forefront of some radical environmental initiatives and has often shown a stronger commitment towards environmental objectives than other German Länder. Bavaria has invested substantially in environmental technology, in particular the monitoring of environmental standards. On the other hand, Bavarian technological perfectionism together with Bavarian self-determination has often resulted in a reluctance to pursue policies from outside the Bavarian *layer*. In addition, the 1990s' economic difficulties and associated problems such as high unemployment have caused Bavarian political-administrative actors and the public to view their (old) environmental policy practices as a luxury that cannot be maintained in the near future. Since the late 1990s, Bavarians have embraced a strategic policy change of deregulation, which now threatens to undermine many Bavarian environmental achievements and indeed many EU environmental policy objectives.

Figure 5.2 highlights formal and informal obstacles in the Bavarian *layer* which make the implementation of EU environmental policies difficult. Bavarian political-administrative actors have scrutinised every EU environmental policy and have relied upon their own media-specific and technological standards. Since the early 1990s, the StMLU in particular has focused on a strategic change towards environmental policy deregulation. Bavaria's deregulation may have been in line with those EU policies which are based on voluntary action. However, other EU environmental objectives have suffered from a lack of discipline and co-ordination. The voluntary approach has been welcomed by the private sector, but environmental NGOs in Bavaria have opposed the new emphasis on voluntary action quite strongly. NGOs have tried to support EU environmental policies but their involvement in the implementation process has been restricted by Bavarian administrators as much as possible. Their confrontational position in Bavarian environmental politics has hindered them from forming a partnership with the public and private sectors which could have supported the implementation of EU environmental objectives.

Figure 5.2: The Bavarian Layer and Environmental Policy

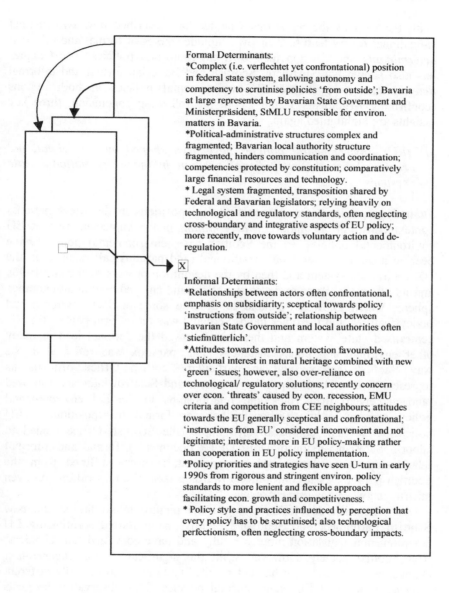

Formal Determinants:
*Complex (i.e. verflechtet yet confrontational) position in federal state system, allowing autonomy and competency to scrutinise policies 'from outside'; Bavaria at large represented by Bavarian State Government and Ministerpräsident, StMLU responsible for environ. matters in Bavaria.
*Political-administrative structures complex and fragmented; Bavarian local authority structure fragmented, hinders communication and coordination; competencies protected by constitution; comparatively large financial resources and technology.
* Legal system fragmented, transposition shared by Federal and Bavarian legislators; relying heavily on technological and regulatory standards, often neglecting cross-boundary and integrative aspects of EU policy; more recently, move towards voluntary action and de-regulation.

X

Informal Determinants:
*Relationships between actors often confrontational, emphasis on subsidiarity; sceptical towards policy 'instructions from outside'; relationship between Bavarian State Government and local authorities often 'stiefmütterlich'.
*Attitudes towards environ. protection favourable, traditional interest in natural heritage combined with new 'green' issues; however, also over-reliance on technological/ regulatory solutions; recently concern over econ. 'threats' caused by econ. recession, EMU criteria and competition from CEE neighbours; attitudes towards the EU generally sceptical and confrontational; 'instructions from EU' considered inconvenient and not legitimate; interested more in EU policy-making rather than cooperation in EU policy implementation.
*Policy priorities and strategies have seen U-turn in early 1990s from rigorous and stringent environ. policy standards to more lenient and flexible approach facilitating econ. growth and competitiveness.
* Policy style and practices influenced by perception that every policy has to be scrutinised; also technological perfectionism, often neglecting cross-boundary impacts.

Conclusion: Subnational Layers and the EU Environmental Policy Process

With the help of the *map*, this Chapter has described how two distinct *sub*national *layers* feature their own 'unique' mixes of formal and informal *determinants* and develop their own environmental policies. The Chapter has also shown how distinctly Scottish and Bavarian formal and informal *determinants* have shaped EU environmental policies in both regions (confirming Arguments 1 and 2). In the following conclusion, three key insights are highlighted again.

(1) (EU) Environmental policies have been shaped by the subnational regions' formal constitutional position in the wider national state systems

*Sub*national regions' formal constitutional positions in the wider national context constitute an important *determinant* in the implementation of EU environmental policies. In the Scottish case, environmental policies have been influenced by Scotland's traditional (and paradoxical) position in the UK centralised system and then by the devolution process of the late 1990s and its aftermath. Prior to devolution, Scotland enjoyed certain autonomous spheres which allowed Scotland to develop some distinct environmental policy features. On the other hand, Scotland was an integral part of the UK centralised state system and therefore followed the national lead in many other environmental policy instances. This paradox was reflected in the way Scottish actors, and in particular Scottish Office officials as representatives of both central government and Scottish interests, followed (and perceived) their roles and relationships in the EU environmental policy process. While responsibilities for formal transposition of EU policies were kept comparatively simple (the Scottish Office tended to adopt a Scottish version of DoE/ DETR documents), formal and informal relationships at a later implementation stage often suffered from the Scottish paradox as it contributed towards a lack of co-operation and even mistrust between government levels.

Scottish devolution changed the picture in so far as the new Scottish Parliament and Scottish Executive have started scrutinising EU environmental legislation more carefully and have consulted 'stakeholders' from Scottish society. However, while this significant formal *determinant* change has taken place, it has not resulted (yet) in a substantially different *filtering* process of EU environmental policies. This observation suggests that formal structures are important but not exclusive in explaining the policy process. Informal *determinants* fill this gap (see below).

Scotland's paradoxical position generated an ongoing debate about devolution that dominated Scottish politics for the past decades. The environmental policy area suffered under this debate in so far as many Scots neglected environmental issues and showed limited interest in the pursuance of EU environmental policies. Nevertheless, environmental problems were not ignored completely and there were indications that many Scots considered 'green' issues increasingly important. Scottish devolution brought an end (at least for the time being) to the public debate over Scotland's political destiny. This development should have encouraged a debate on an independent and meaningful environmental policy in a devolved Scotland. However, despite pressures from 'green' NGOs and other organisations,[129] this 'historic opportunity' has not been taken up fully to this day. Other policy priorities (jobs, transport, social security) prevail in Scotland and put a damper on environmental objectives. It can be deduced that while formal structures may have changed, informal attitudes towards the environment have not changed significantly. It may be too early at this stage to determine the impacts of Scottish devolution on the *filtering* of EU environmental policies in the Scottish *layer*. At this stage it appears that the net difference between policy implementation before and after Scottish devolution is minimal.

In contrast to Scotland, Bavaria has enjoyed for a long time substantial political powers in the federal state system. Bavarians have therefore been able to establish their own independent environmental policy. Arguably, Bavaria has been the German Land most committed to the environment and has stood at the forefront with a number of environmental initiatives. Until the mid 1990s, Bavaria managed to accommodate traditional values and 'new' environmental concerns (such as nuclear safety and air pollution) as matters of central importance. However, with similar vigorousness, Bavaria has conducted an environmental policy 'U-turn' in recent years. The new lenient and flexible approach has proven to be more substantial and radical than the environmental policy adjustments of Germany as a whole (see Chapter 4).

Bavaria's relative independence in the environmental policy area has not left much room for environmental ideas and instructions from 'outside'. In fact, every EU environmental policy that entered Bavaria's fragmented and legalistic political-administrative system has faced serious implementation difficulties. Bavarian political-administrative actors have protected and maintained their competencies and have demonstrated a rather confrontational attitude towards instructions from 'above' or 'outside'. This sceptical and confrontational attitude has meant that EU environmental policies have had to overcome (and sometimes have failed) constitutionally protected hurdles of Bavarian scrutiny and approval. The case of Bavaria confirms Argument 1; formal and informal *determinants* are inter-related and their 'combined force' influences the implementation of EU environmental policies.

(2) (EU) Environmental policies have been influenced by distinctly Scottish and Bavarian Informal Determinants

While the above formal *determinants* are important, it would be misleading to place too much emphasis on Scotland and Bavaria's constitutional positions and internal formal structures. As indicated above, there are informal *determinants* such as policy-makers' priorities, attitudes towards environmental protection, and policy styles within the *subnational layers* which ultimately shape EU environmental policies. Scotland has been, in some respects, 'greener' than the UK as a whole: Scotland's unique natural environment has been an integral and much valued part of Scotland's national heritage. In addition, important economic sectors such as tourism and the wool and whisky industries have depended upon a healthy environment in Scotland. Therefore, commercial considerations have contributed towards an interest in environmental matters in Scotland. On the other hand, there are indications that the majority of Scots have been, and still are, less concerned about environmental problems (such as water pollution) than other EU citizens. In the past this half-heartedness could be explained with the Scots' devolution debate. Another explanation could (and can) be found in Scotland's abundance of natural resources, which make environmental problems less visible. A more plausible explanation lies in economic priorities. Because of Scotland's peripheral situation these priorities have been perceived as more pronounced and urgent than in other parts of the EU. While economic priorities continue to distract many key Scottish actors from environmental issues and the implementation of EU environmental policies, recent years have also seen a slight upward trend in the Scots' environmental awareness. This new 'green' interest can only benefit EU environmental policy implementation in the near future, but it has not been reflected in the *filtering* of these policies up until now.

Informal *determinant* differences are also evident between Bavaria and the Federal level: Bavaria was the first German Land to formulate an environmental policy. The Bavarians' motivation to pursue an environmental policy as early as 1970 derived mainly from Bavaria's Flächenstaat characteristics (large rural, agricultural and forestry areas) and the traditional values associated with them, as well as the desire to promote economic development without compromising Bavaria's natural resources to an unacceptable level. While Bavarians combined successfully traditional and 'new' environmental values, they have also reacted to the economic challenges of the 1990s with a policy change more radical than in the rest of Germany. The Bavarians' perceptions and attitudes, which have obviously changed over time and have depended on political-economic circumstances, have influenced the way in which EU environmental policies have been processed. Since the early 1990s, Bavarians have pursued a policy of substitution and deregulation which is compatible with voluntary action tools of the EU. Bavaria's new strategy, however, is less compatible with regulatory and procedural EU environmental policy obligations. While Bavarian political-administrative actors have reassured critics that deregulation will not lead to lax environmental policy compliance, the risk still remains that implementors and practitioners will not follow EU policy requirements by the book.

(3) Formal and informal links between the EU and subnational regions have influenced significantly EU environmental policy implementation

Finally, the *sub*national regions' formal and informal links with the EU have influenced EU environmental implementation performances. Both Scottish and Bavarian links with the EU have varied. While the Bavarians have enjoyed stronger formal links with Brussels and have had the financial and administrative means to channel their views to the EU level, the Scots have participated either through the national *layer* or through well-functioning informal lobbying bodies in Brussels. Scottish devolution has not changed this set up, as the EU remains a 'reserved matter' for Westminster and Whitehall. Bavarians, on the other hand, have campaigned for, and have maintained, strong formal links with the EU in order to influence the formulation of EU policies. Indeed, their rigorous campaign for more participation has occasionally backfired, as EU actors have often perceived Bavarian lobbying as intimidating. Bavarians have been less interested in close ties with the EU (Commission) when it came to the implementation of EU environmental policies. This discrepancy suggests that Bavarians are not particularly interested in the success of EU environmental policies. Although this may be true to a certain extent, the main reason for this discrepancy lies in the Bavarians' confidence that they fully comply with EU obligations. Another reason can be found in their determination not to be supervised by another government level. This attitude, however, prevents any critical self-assessment on the part of Bavarian implementors which, ultimately, would be beneficial for EU environmental objectives.

The Scots have appeared to be more 'euro-cooperative' than their Bavarian counterparts. Until the Labour victory of May 1997, Scottish local government officials in particular have considered the Commission as an ally against the 'mighty' central (Conservative) government and have therefore participated at the EU level in a less confrontational manner than the Bavarians. This 'by-passing' of Westminster and Whitehall may have mellowed with the new Labour Government, but it may well reoccur in another form in future if Westminster and Edinburgh (and Scottish local governments) clash over EU issues that affect several government levels.

Having highlighted Scotland's cooperative approach towards Brussels, a 'truly euro-friendly' position is nevertheless difficult to establish. Firstly, Scottish links with Brussels have been maintained through central government and (opaque) informal channels. Disagreements between Scottish and EU levels have therefore been less obvious. Secondly, Scots have tended to more 'euro-sceptic' when faced with the implementation of EU environmental policies. Confronted with 'inconvenient' practicalities of EU environmental policies, many Scots (particularly those affected by 'unbearable' administrative and economic costs imposed on their already ailing economies) have turned out to be as 'euro-sceptical' as their Bavarian colleagues. Therefore, the implementation of EU environmental policies in Scotland has not been guaranteed by the Scots' apparent 'euro-friendliness'; instead it has depended upon Scottish pragmatism and economic (self-) interests in a competitive Single Market.

The difference in EU attitudes between *sub*national and national *layers* is striking. In contrast to the German 'driving force' in the European integration process, Bavarians have taken an overall 'euro-sceptical' stance and have defended their powers and interests at every opportunity. The Scots, on the other hand, have departed from the UK position of an openly 'awkward EU partner'[130] and have presented themselves as a 'euro-friendly' partner. This image, however, has not prevented occasional criticism against 'bright-eyed bushy-tailed Commission officials'[131] by Scottish actors who have faced 'inconvenient' EU environmental obligations that placed a disproportionally heavy burden on the peripheral region Scotland.

Looking at the overall process of EU environmental policy implementation, the question remains whether EU environmental policies are compatible with the Scottish and Bavarian political-administrative systems. If both *sub*national *layers* feature *determinants* which are favourable to EU environmental policies, policy implementation should take place without major difficulties. However, if *sub*national *determinants* are incompatible with EU environmental policies, problems of implementation are almost inevitable. Focusing on formal and informal links between government levels, both Scotland and Bavaria have featured unfavourable weaknesses and gaps in the EU environmental policy *filtering* process. Tensions caused by political and constitutional circumstances exist in both political-administrative systems. In the Scottish case, vertical communication and co-operation links have been for most part dominated by mistrust and political differences (especially between the former Scottish Office and Scottish local authorities). Horizontal links between policy sectors have been more integrated in Scotland but have also shown problems in co-ordination (especially within local authority administrations). The UK centralised state system implied one advantage for the implementation of EU environmental policies: since the Scottish Office and the Secretary of State were part of UK central government and since Scotland did not possess a parliament and government of its own, policy instructions did not require political approval in the *sub*national *layer*. Scottish devolution introduced a formal structure that now allows Scots to properly scrutinise (and shape) EU policies before they are implemented in Scotland. While this added formal structure may have eliminated the earlier mentioned advantage in the *filtering* process, it has not (yet) led to substantial changes in attitudes towards EU environmental policies nor has it affected significantly the actual implementation of these policies.

In contrast, the transposition of EU environmental policies in Bavaria was bound to be difficult from the start. The Federal German constitution (and the Bavarian constitution for that matter) emphasises the sharing of competencies between government levels. While Bavarian implementors are required to follow their obligations under the Federal constitution and under the Treaties of the EU, they also consider it legitimate to assess and shape every policy which enters their 'territory'. Their interpretation and 'additions' to policies 'from above', often in the form of technological and legalistic details, can result in the (arguably) inadequate implementation of EU environmental policy objectives. In this respect, Bavaria appears to be in a disadvantage as far as the German

'Politikverflechtung' is concerned (see also Chapter 4). In addition, the Federal and Bavarian political-administrative systems and their environmental policies in particular are compartmentalised and fragmented. While horizontal fragmentation ensures a certain amount of independence for environmental actors to pursue ambitious policies, it also has the effect that integrative EU environmental policies such as the EIA Directive are not co-ordinated and implemented properly. Despite these gaps, one Bavarian *determinant* is favourable for the pursuance of EU environmental policies: Bavaria possesses the necessary financial and administrative resources (although perhaps less impressive in the 1990s) as well as technological know-how in the environmental field and should therefore be able to meet EU environmental policy obligations even if EU qualitative and quantitative standards are different from Bavarian standards.

This Chapter has highlighted the Scottish and Bavarian *layers* and assessed to what extent their formal and informal *determinants* have shaped EU environmental policies on the ground. The following case study on the implementation of the EIA Directive (85/337/EEC) examines the Scottish and Bavarian *layers* and their experiences with the policy. In order to highlight more clearly the discrepancy between national and *sub*national *layers*, the case study compares and contrasts Scottish and Bavarian performances with the implementation findings of the UK and Germany at large.

Notes

[1] Other aspects in which Scotland retained its national identity include culture, the media and sport. For a discussion on Scotland's position in the UK see Kellas (1989) and Midwinter et al (1991).
[2] Helen Liddell is the first woman Secretary of State for Scotland. She was appointed in January 2001.
[3] The Secretary of State for Scotland is supported by a new Scotland Office based in Whitehall. In recent months, however, the media and some politicians have expressed their doubts whether the position of Secretary of State in Westminster should be maintained given that 'Scots now look after themselves'.
[4] Quoted in Butt Philip and Baron (1988).
[5] Interview, 30. May 1995, Edinburgh.
[6] Some critics of devolution (particularly members of both the SNP and the Conservative Party) have suggested that links will become weaker, with the result that Scotland may become more (or 'too') insular. For a detailed account of actors and their views and expectations regarding Scottish devolution and its impact on links with Europe, see

Sloat (2001).

[7] At the general election in 1992 three quarters of the Scottish electorate voted for opposition parties - the Labour Party, the Liberal Democrats and the Scottish National Party.

[8] The Council Tax and the creation of 'quangos' in Scotland serve as examples.

[9] In particular the UK Deputy Representative, a Scot. Information from UKREP official, interview, 5. March 1997, Brussels.

[10] A UKREP official described the mistrust between Scottish and UK officials in an interview, 5. March 1997, Brussels.

[11] The Scottish Executive European Office, part of Scotland House, works 'closely together' with UKREP, according to the office's home website (www.scotland.gov.uk/euoffice). While views differ to what extent UKREP and Scotland House will clash in future over UK and Scottish interests, most observers and actors acknowledge that there is a potential (or 'possibility') for conflict. See Sloat (2001).

[12] Many Scots considered EU actors and institutions such as the Commission as allies in their campaign for devolved powers for Scotland and used EU links to 'bypass' Westminster.

[13] Among them CoSLA, East of Scotland European Consortium (ESEC), Edinburgh's Telford College, Eurodesk Brussels Link, Highlands & Islands of Scotland European Office, Maclay Murray & Spens (law firm), Scottish Enterprise, West of Scotland European Consortium.

[14] Eight MEPs represent Scotland; among them David Martin (EP Vice-President).

[15] For another account of Scottish (pre-devolution) representation at the EU level see Bomberg (1994).

[16] The Scottish Executive EU Office in Brussels has currently seven members of staff and two trainees.

[17] This complementary role was stressed by the then First Minister for Scotland, Donald Dewar, in his speech to MEPs, 9. February 1999.

[18] See Himsworth and Munro (1998) for further details.

[19] Examples where Westminster and Edinburgh have differed already include provisions for pensions, tuition fees and terms and conditions of service for teachers.

[20] According to 1998 opinion polls conducted in Scotland by MORI, System Three and others, SNP support for the next Scottish Parliament elections has been almost equal to the support of the governing Labour Party. See *The Scotsman* (1998).

[21] See McDowell (1995).

[22] 'The Scottish Parliament offers an opportunity to start afresh and learn from not just the UK experience, but from our sister countries in Europe, some of whom have a longer and more thoroughgoing approach and strategy from which we can learn.' John Wheatley Centre (1997).

[23] Scottish Office official, telephone interview, 5. April 1995.

[24] A similar observation was presented by Sloat (2001): positions and policy implementation did not differ 'in substance' but more in 'emphasis'.

[25] See Scottish Parliament (2001) press release.

[26] Scottish Office official, interview, 30. May 1995, Edinburgh.

[27] For instance, the Scottish Office, which was responsible for setting up the Scottish Environmental Protection Agency, was blamed by environmental groups for not including 'integrated catchment management' (ICM) as one of SEPA's responsibilities. Among other aspects, ICM takes into account the effects of intensive farming and

intensive forestry methods on water resources. Environmental groups claim that ICM was left out deliberately to please the farming community.

[28] Green MSP Robin Harper expressed these views in an article in *The Press and Journal* (March 2001).

[29] The Earl of Lindsay initiated a Scottish Office publication documenting environmental initiatives. *Common Sense, Common Purpose. Sustainable Development in Scotland 1996* is a colourful yet superficial statement. Lord Sewel was committed 'to place sustainable development at the heart of policy making'. Quoted from *The Herald* (1997).

[30] See Scottish Executive press release (SE2846), 3. November 2000.

[31] Having moved from (New) St Andrew House, Victoria Quay is the new Scottish Executive Headquarters in Leith, Edinburgh.

[32] Regional Council official, interview, 11. January 1995, Edinburgh. This situation has changed recently, following the Labour Government White Paper on Transport of 1998, which proposed more discretionary powers for local authorities. New powers include charging motorists for the use of roads and parking. See *The Economist* (1998).

[33] These views were expressed by a Council official during a public lecture, held at Glasgow University, 7. March 2001.

[34] See *The Herald* (1997).

[35] For example, Local Agenda 21 plans had to be abandoned at the time of local government reform. Contacts with environmental actors could not be maintained during the shake-up. CoSLA spokesperson for Local Agenda 21, interview, 25. April 1997, Stirling.

[36] Examples include *Environmental Action in Strathclyde* (1994), *Environmental Strategy* (Edinburgh, 1995), *Charter for Action on the Environment* (Lothian, 1994) and *Environmental Charter* (Central Region, 1994).

[37] Former Lothian Regional Council, for example, promoted and financially supported LEEP (Lothian and Edinburgh Environmental Partnership) which advised businesses and citizens on recycling and energy efficiency.

[38] Regional Council officials, interviews, 11. January 1995 (Edinburgh) and 7. August 1995 (Glasgow).

[39] In fact, Scottish local authorities prepared a 'Good Practice Guide' on sustainable development (together with CoSLA) and have recently met at a conference 'Sustaining a Nation, Scotland + 10' (March 2001) to develop a common Local Agenda 21 approach. The aim of this conference was to formulate a joint Scottish document for the forthcoming follow-up conference of the Rio Earth Summit, which will be held in South Africa in 2002.

[40] There have been variations, of course; some environmental officers have been more influential than others. Overall, however, it can be said that Scottish local authorities have not fully integrated environmental considerations into all aspects of local government policy.

[41] Former Regional Council environmental officer, interview, 11. January 1995, Edinburgh.

[42] One indication for the lack of interest is the Scots' poor performance in recycling waste; Scots were at the bottom of the European-wide league table for collecting and recycling waste in 2000. See the Accounts Commission for Scotland survey results for the year 1999/ 2000. www.audit-scotland.org.

[43] Description used by Scotland Europa Ltd Spokesperson, interview, 5. March 1997, Brussels.

[44] See McCaig and Henderson (1995).

[45] The dumping plan caused boycott protests throughout Europe with an immediate 20% decrease in Shell's retail figure in Germany. In contrast, car drivers in Scotland continued to buy their petrol at Shell petrol stations, according to a BBC news report (1997).

[46] Baker et al (1994) confirm the peripheral regions' concern that so-called 'core regions' are more economically advanced and that they enjoy better trade links with the rest of the EU.

[47] Bomberg (1994) describes Scotland's 'peripherality' and gives examples on the (perceived) disproportional costs.

[48] Information from UKREP official. Interview, 5. March 1997, Brussels.

[49] The comment was made by a representative from the Scottish whiskey industry at an Environmental Group meeting organised by Scotland Europa. Glasgow, 26. February 1997.

[50] The comment was made by a Scottish business representative at an Environmental Group meeting organised by Scotland Europa, Glasgow, 26. February 1997.

[51] Prior to 1996 these functions belonged to Her Majesty's Industrial Pollution Inspectorate (HMIPI) and the River Purification Boards (RPBs), as well as Scottish local authorities.

[52] ICM (integrated catchment management) takes into account the effects of intensive farming and intensive forestry methods on water resources.

[53] Examples include an 'Environmental Law Lecture' at the University of Edinburgh (15. January 1997) and a Scotland Europa Environmental Group meeting (26. February 1997).

[54] SEPA is going to play a central role in the implementation of the new 'water framework' Directive (2000/60/EC).

[55] In fact, SNH is financially dependent on government funding. Quotations from SNH (1994, 1995 and 1997).

[56] Examples include the SNH's 'Countryside around Towns' programme, the 'Coastal and Marine Task Force' and the 'Cairngorms Project'.

[57] SNH prepared a list of areas for environmental protection which fall under the Habitats Directive.

[58] In 1995 FoEScotland informed the Commission about the M77 road extension project which caused controversy and media attention in Scotland. Commission DG XI replied promptly and sympathetically to FoEScotland's claims that the M77 project did not comply with requirements of the EIA Directive. FoEScotland spokesperson, interview, 17. April 1995, Edinburgh.

[59] FoEScotland spokesperson, interview, 17. April 1995, Edinburgh.

[60] REEFs stands for 'Regional Environmental Education Forums'. The SEEC also collaborated with many other institutions concerned with environmental education such as teacher training colleges and the Scottish Office.

[61] With the binder *Environmental legislation and policy for the manager*, CEBIS regularly updated Scottish businesses about new standards and requirements in the environmental field.

[62] For a discussion on a future Scottish environmental policy see John Wheatley Centre (1997).

[63] Environmental activists have already expressed their fears that a future Scottish Parliament will focus on 'jobs at all costs' while neglecting the environment. See FoEScotland (1998).

[64] For a general introduction to the politics of the Free State of Bavaria see James (1995).

[65] The Federal German constitution provides a clear framework: in policy areas such as defence the Federal level enjoys exclusive decision-making powers (Art.73), 'concurrent' or shared powers exist in areas which concern both government levels such as social services (Art.74), and in areas such as education the Federal level only provides framework guidance while respecting the 'cultural diversity' of the Länder (Art.75).

[66] A German COREPER official described the Bavarian administration as a 'Musterverwaltung', a perfectionist 'model' administration which is not prepared to accept instructions from outside. Interview, 5. March 1997, Brussels.

[67] The Senate has no counterpart in any of the other German Länder. It consists of sixty members representing social, economic, cultural and local community interests who must be older than 40 years of age. Among other sources, for more information about policy-making in Bavaria see www.bayern.landtag.de\wissen\gesetz\gesetz.htm.

[68] To date, five referenda have been held in Bavaria. One referendum concerned the question whether Bavaria should align its electoral system to the Federal system. Following the 1970 referendum, Bavaria changed its '10% electoral hurdle' to the Federal '5% hurdle'. Most recent referendum 1. October 1995: 'Volksentscheid über neue kommunale Mitwirkungs- und Entscheidungsrechte der Bürgerinnen und Bürger' (referendum on citizens' participatory powers in local authority decision-making).

[69] Apart from the Bayerische Staatskanzlei, the other nine government departments are: Staatsministerium für Bundesangelegenheiten (federal matters); Bayerisches Staatsministerium (B.S.) des Innern (interior); B.S. der Justiz (justice); B.S. für Unterricht, Kultus, Wissenschaft und Kunst (education, culture, science and arts); B.S. der Finanzen (finances); B.S. für Wirtschaft, Verkehr und Technologie (trade and industry, transport, technology); B.S. für Ernährung, Landwirtschaft und Forsten (food safety, agriculture and forestry); B.S. für Arbeit und Sozialordnung, Familie, Frauen und Gesundheit (employment, social security, family, women and health); and the Bayerische Staatsministerium für Landesentwicklung und Umweltfragen (planning, development and environmental matters).

[70] To date, Bavaria has not seen a woman Ministerpräsident.

[71] For an empirical analysis of the Bavarians' unusual electoral behaviour see Falter (1982). Bavarian elections in September 1998 confirmed that the Bavarian electorate differs from the rest of Germany: the CSU secured an absolute majority while the general elections conducted one week later resulted in a red-green coalition government at the Federal level.

[72] For example, (late) Franz Josef Strauß' regular disputes with (former) Federal Chancellor Helmut Kohl. More recently, Bavarian Environment Minister Schnappauf (CSU) attacked Federal Environment Minister Trittin (Die Grünen) on a number of issues, including EU environmental policy compliance and the phasing out of nuclear energy. An example for Länder quarrels include Stoiber's criticism in 1997 over the 'lax' handling of anti-nuclear protests by the Lower Saxony police force.

[73] For a Bavarian critique of EU environmental 'instructions' see in particular Wegner (1993). More recently Schnappauf renewed the Bavarian criticism of 'another gigantic wave of Eurocracy approaching Bavaria'. See StMLU website press release.

[74] Unterfranken, Oberfranken, Mittelfranken, Oberpfalz, Niederbayern, Oberbayern, Schwaben.

[75] The Counties themselves are divided into smaller communities, totalling 2052 Gemeinden in Bavaria. District presidents are appointed civil servants; members of the District

assemblies (Bezirkstage) are elected. County presidents (Landräte) as well as County Councils (Kreistage) are elected by their constituents; the same applies to Town Councils, town mayors, and community Councils (Gemeinderäte).

[76] Spitzenverbände in Bavaria: Verband der Bayerischen Bezirke (regions), Bayerischer Landkreistag (districts), Bayerischer Städtetag (towns), Bayerischer Gemeindetag (municipalities).

[77] The following statement indicates that Bavarian local authorities have to guard their constitutional rights: the Städtetag pointed out that 'constitutional reality is often not identical with constitutional right'. (in German: Denn Verfassungsanspruch und Verfassungsrealität driften immer wieder auseinander.) See Bayerischer Städtetag (1991).

[78] Stiefmütterlich translated into English: 'behaving like a stepmother'. Comment made by a former official of the Europabüro der Bayerischen Kommunen, interview, 6. March 1997, Brussels.

[79] For details on EMU convergence criteria see Chapter 4.

[80] For instance, some social security provisions and environmental policy tasks were shifted to local authorities.

[81] Official of the Europabüro der Bayerischen Kommunen, interview, 4. March 1997, Brussels.

[82] Länder representatives met in 1987 to adopt the 'Ten Munich Theses on European Policy'; Bavaria initiated and hosted the first Conference on 'Europe of the Regions' in 1989.

[83] The four key demands of the regions: formalisation of the subsidiarity principle; *sub*national representation at Council meetings; a regional chamber; a right of appeal for *sub*national governments to the European Court of Justice. Apart from the latter, all demands were met with the Maastricht Treaty of 1992.

[84] Article 23 was included in the constitution without, however, amending Articles 24(1) and 32(1).

[85] Länder representatives participate on behalf of the Federal Republic of Germany as a whole. Article 23 is further strengthened by a more detailed Federal Law of 12 March 1993 and an Agreement on the Co-operation in European Union Matters of 29 October 1993. Prior to these changes, the Federal German constitution provided the Federal level with exclusive decision-making powers at the EU level; the Länder could only observe developments from a distance (the only Länder official in Brussels was the 'Beobachter der Länder' ('observer') whose tasks were restricted to the collection of EC information for the Länder).

[86] According to a Bavarian Government official, the current situation is 'not satisfactory': Länder can send delegates to the Council but still have no independent representation. Written correspondence, 13. March 1996, Munich.

[87] Terms used by Rometsch (1995).

[88] Bavarian interests are also pursued by Members of the European Parliament, Members of the Committee of the Regions, and, of course, Bavarians who work within EU institutions. A similar index to 'Jock Tamson's Bairns' does not exist for Bavaria, but individual contacts are maintained on an informal basis.

[89] Generally, German MEPs do not represent constituencies as such due to the mixed FPTP/PR electoral system. However, fourteen MEPs are elected as Bavarian members under the FPTP section. Information from *Europa-Pass für Bayern* (no date).

[90] For instance, a Commission DG XI official complained about the 'most severe

intervention' (massivste Intervention) tactics by Bavarian representatives during the preparation of an environmental policy proposal. Commission DG XI official, interview, 7. March 1997, Brussels.

91 Constitutional reform in 1984. Article 3(2) commits the Free State to protect the natural resources and cultural traditions; Article 131(2) establishes a 'sense of responsibility towards the environment' as one of the key objectives in Bavarian education; Article 141(1) obliges the Free State and its citizens to protect the environment in a sustainable manner for future generations. For a detailed account of Bavaria's development and adoption of the environmental objective see Mauritz (1995). For a 'green' NGO (critical) assessment as to whether the Bavarian State government lives up to the environmental provisions, see Bund press release (2000).

92 Since 1994, the Federal German constitution includes a new Article 20A: the state protects the environment for future generations in accordance with the constitution and legislation.

93 This principle, however, is limited to cases where economic and other considerations have 'long-term and fundamental impacts'.

94 For further information see StMLU (1994 and 1996) and LfU (no date).

95 English translation: Bavarian State Ministry for Planning, Development and Environmental Matters.

96 Bavarian political-administrative actors thereby responded to the negative impacts on the environment of post-war economic reconstruction and migration from Central and Eastern Europe, both of which were completed by the late 1960s.

97 One StMLU official commented that other government departments and local governments do not report to him and his colleagues on EIA policy compliance. Interview, 19. August 1996, Munich.

98 See Bund Naturschutz (1993).

99 'Ist zu' (1976 Programme) was replaced by 'soll', 'möglichst' and 'soweit möglich und vertretbar' (1993 Programme). See Bund Naturschutz (1993a).

100 See StMLU's *Umweltpakt Bayern* (1995). English translation: pact on the environment.

101 For a summary of the 'substitution and deregulation' approach see Böhm-Amtmann 1998).

102 See Free State Government and StMLU websites for more information.

103 See www.bund-naturschutz.de for further information.

104 These views were openly expressed during a research interview. StMLU officials, 19. August 1996, Munich. Further, see Wegner (1993). More recently, the StMLU complained that the EU and the Commission in particular ignored the principle of subsidiarity and 'imposed unnecessary and inconsistent environmental regulations'. These regulations are, according to the StMLU, 'counterproductive'. However, the 6th EAP signalled that the Commission 'was coming round' to the Bavarian approach. See StMLU (2001).

105 StMLU written correspondence, 26. February 1998, Munich.

106 Neither the Treaties nor the Federal German constitution provide for specific measures which monitor and control implementation performances within the Länder and Bavaria in particular.

107 See StMLU (2001a).

108 For instance, Thomas Goppel (former Bavarian State Government Minister for the Environment) opposed the EU 'dirigism' during a speech at the Peutinger-Collegium, 23. April 1997, Munich.

[109] One exception is, perhaps, Thuringia which followed Bavaria's 'Euro-sceptic' example. The Thuringia Land government clashed with the Commission over subsidies for declining industries in Thuringia which the Commission considered as distorting the Single Market.

[110] The Bavarians' scepticism does not stop at the EU (Commission) level: apart from complaints by (former) StMLU Minister Thomas Goppel about the Commission's inactivity in monitoring regions' performances, other EU regions have been criticised for not complying with EU environmental legislation (Press Release PM-Nr.126/97). Goppel's remarks as well as strong critical views expressed to the author by StMLU officials indicate that Bavarian politicians and administrators are not inclined to adjust their informal attitudes in order to co-operate more effectively with the Commission on the implementation of EU environmental policies (interview, 19. August 1996, Munich).

[111] According to an official of the Europabüro der Bayerischen Kommunen, 4. March 17, Brussels.

[112] The survey *Die umweltbewusste Gemeinde* was conducted by B.A.U.M Consult München GmbH. Findings were being assessed at the time of writing.

[113] KommA21 in full: Kommunale Agenda 21 Bayern. For more information see Free State Government website.

[114] Examples include *Informationen zu Naturschutz und Landschaftpflege* (Oberbayern, no date).

[115] Examples include *Umweltschutz im Regierungsbezirk Unterfranken* (1996) and *Umweltschutz im Regierungsbezirk Mittelfranken* (1995).

[116] Examples include *Landschaftsplanung Gemeinde Kirchdorf im Wald* (no date), *Gemeinde Hunding Arbeitsergebnisse: Kommunale Strukturpolitik* (1995) and *Bürgerinformation Landschaftsplan der Stadt Abensberg* (no date).

[117] 88% of Bavaria's land mass is used for forestry and agriculture. For more information see Hinterstoisser (1996).

[118] English translation: Federation for Environmental Protection in Bavaria (e.V. abbreviation for 'eingetragener Verein' - registered organisation). Bund Naturschutz in Bayern has 130,000 members and is associated with the German-wide BUND (Bund für Umwelt und Naturschutz).

[119] See Bund Naturschutz (1995).

[120] See Bund Naturschutz (1993 and 1989).

[121] English translation: Federation for the Protection of Birds in Bavaria.

[122] Information provided by two Bund activists during research interview, 19. August 1996, Landshut (Bavaria).

[123] The list had already been published by the time the LBV was consulted for final comments. Information from LBV written correspondence, 2. February 1998, and LBV press release (1997).

[124] For further information see LfU (no date) and StMLU (1996b).

[125] Of the EU GDP average 100, Bavaria scores 127. Bavaria's wealth has been one of the key reasons for Bavaria's strong environmental policy stance, according to a representative of the Bavarian local authorities office in Brussels. Interview, 4.March 1997.

[126] The policy of derogation and deregulation is outlined in StMLU (10/96) and the *Umweltpakt Bayern* (1995).

[127] Wegner (1993) highlights incompatibilities between EU and Bavarian standards and describes many EU environmental policies as 'Rückschritte' ('backward steps') for

Bavaria.
[128] Bavarian StMLU official, interview, 19. August 1996, Munich.
[129] See John Wheatley Centre (1997), mentioned above. FoEScotland (1997) urged environmentalists to keep up the pressure and remind politicians in Scotland that environmental concerns should be considered in a (future) Scottish Parliament.
[130] George (1990) uses this description.
[131] The comment was made by a representative of the Scottish Whisky industry, see above.

References

Baker, S. et al (1994), *Protecting the Periphery. Environmental Policy in Peripheral Regions of the European Union*, Frank Cass, London.
Bayerische Staatskanzlei (1995), *Umweltpakt Bayern. Miteinander die Umwelt schützen.*
Bayerischer Städtetag (1991), *Aufgaben, Organisation, Mitglieder.*
BBC (1997), *Scotland Today*, TV news report, (16. June).
Böhmer-Amtmann, E. (1998), *Cooperative Enforcement: Successes, Impediments, Solutions*, Conference Paper at the United States Environmental Protection Agency, Washington, (21-22. January).
Bomberg, E. (1994), 'Policy Networks on the Periphery: EU Environmental Policy and Scotland', *Regional Politics and Policy*, vol.4, No.1, pp.45-61.
Bund Naturschutz in Bayern e.V. (1989), *BN-Position. Das Alpenprogramm des Bundes Naturschutz.*
Bund Naturschutz in Bayern e.V. (1993), *Rettet die Donau! Stoppt die Kanalisierung!*, leaflet, (December).
Bund Naturschutz in Bayern e.V. (1993a), *BN-Position. Ökologisches Landessanierungsprogramm. Stellungnahme zur Fortschreibung des Landesentwicklungsprogramms Bayern.*
Bund Naturschutz in Bayern e.V. (1995), *BN-Position. Zukunft für die Landwirtschaft. Aktualisierte agrarpolitische Forderungen des BN*, (November).
Bund Naturschutz in Bayern e.V. (2000), *Bund Naturschutz bekräftigt Unterstützung des Volksbegehrens 'Macht braucht Kontrolle'*, press release PM14/LGS, (May).
Butt Philip, A. and Baron, C. (1988) 'UK Report', in H. Siedentopf and J. Ziller (eds), *Making European Policies Work. The Implementation of Community Legislation in the Member States*, Sage, London.
CEBIS (no date), *Environmental legislation and policy for the manager.*
Central Regional Council (1994), *Environmental Charter.*
City of Edinburgh District Council (1995), *Environmental Strategy*, Department of Strategic Services.
Falter, J. (1982), 'Bayerns Uhren gehen wirklich anders. Politische Verhaltens- und Einstellungsunterschiede zwischen Bayern und dem Rest der Bundesrepublik', *Zeitschrift für Parlamentsfragen*, Nr.13, pp.504-521.
FoEScotland (1997), 'Scottish Parliament - Green for Go?', *What on Earth*, Issue 18, Summer, p.6.
FoEScotland (1998), 'View from the Mill', *What on Earth*, Issue 20, Spring, p.3.
Gemeinde Kirchdorf im Wald (no date), *Landschaftsplanung Gemeinde Kirchdorf im Wald.*

Gemeinde Hunding (1995), *Gemeinde Hunding Arbeitsergebnisse: Kommunale Strukturpolitik.*
George, S. (1990), *An Awkward Partner: Britain in the European Community*, Clarendon Press, Oxford.
Harper, Robin (2001), 'Spreading the environment too thinly', *Press and Journal*, (23. March).
Himsworth, C. and Munro, C. (1998), *Devolution and the Scotland Bill*, W. Green and Son, Edinburgh.
Hinterstoisser, F. (1996), 'Umweltpolitik in Bayern - Folgen für die Landwirtschaft', *Landwirtschaft und Umweltpolitik*, Nr.30, pp.25-33.
James, P. (1995), *The Politics of Bavaria - An Exception to the Rule*, Avebury, Aldershot.
John Wheatley Centre (1997), *Working for Sustainability: An Environmental Agenda for a Scottish Parliament*, Final Report, The Governance of Scotland Project, Edinburgh.
Kellas, J. (1989), *The Scottish Political System*, Cambridge University Press, Cambridge.
LBV (1997), *Landesbund für Vogelschutz (LBV) fordert: Schutzgebietnetz NATURA 2000 erweitern*, Presseinformation Nr. A31-97, (22. July).
LfU (no date), *Im Dienste des Umweltschutzes.*
Lothian Regional Council (1994), *Charter for Action on the Environment. 3rd Environmental Action Plan.*
Mauritz, M. (1995), *Natur und Politik: Die Politisierung des Umweltschutzes in Bayern*, Andreas Dick Verlag, Neustraubling.
McCaig, E. and Henderson, C. (1995), *Sustainable Development: What it means to the general public*, MVA Consultancy survey commissioned by the Scottish Office Central Research Unit.
McDowell, E. (1995), 'The Environmental Movement in Scotland: An Empirical Analysis of Leading Group Activists', in J. Lovendusky and J. Stanyer (eds), *Contemporary Political Studies*, vol.3, Political Studies Association, University of York.
Midwinter, A., Keating, M. and Mitchell, J. (1991), *Politics and Public Policy in Scotland*, MacMillan, London.
Regierung Oberbayern (no date), *Informationen zu Naturschutz und Landschaftpflege*, newsletter.
Regierung Mittelfranken (1995), *Umweltschutz im Regierungsbezirk Mittelfranken.*
Regierung Unterfranken (1996), *Umweltschutz im Regierungsbezirk Unterfranken.*
Rometsch, D. (1995), *The Federal Republic of Germany and the European Union. Patterns of Institutional and Administrative Interaction*, University of Birmingham Discussion Papers in German Studies.
Scottish Executive (2000), press release (SE2846), (3. November).
Scottish Office and CoSLA (1993), *Local Environment Charter for Scotland*, (September).
Scottish Natural Heritage (1994), *Third Operational Plan - Review of 1993-94.*
Scottish Natural Heritage (1995), *Work Programme 1994-95.*
Scottish Natural Heritage (1997), *Corporate Plan 1994\95 - 1996\97.*
Scottish Parliament (2001), *Transport and Environment Committee publishes GMU Report*, press release (CTRAN01/2001), (23. January).
Sloat, A. (2001), *Scotland in the European Union: Expectations of the Scottish Parliament's Architects, Builders, and Tenants*, Scotland Europa Paper 22, (March).
Stadt Abensberg (no date), *Bürgerinformation Landschaftsplan der Stadt Abensberg.*
StMLU (1994), *Umweltschutz in Bayern '94.*
StMLU (1996), *Information: Umweltschutz und Landesentwicklung in Bayern.*

StMLU (1996a), *Die Umweltbewusste Gemeinde. Leitfaden für eine nachhaltige Kommunalentwicklung.*

StMLU (1996b), *Umweltschutz und Landesentwicklung in Bayern.*

StMLU (1999), *Plan-UVP verursacht neue Prüfpflichten, mehr Zeit- und Kostenaufwand,* press release, www.umweltministerium.bayern.de.

StMLU (2001), *Umweltschutz in der Europäischen Union – Anspruch und Wirklichkeit,* www.umweltministerium.bayern.de.

StMLU (2001a), *Trittin operiert mit falschen Zahlen,* press release, www.umweltministerium.bayern.de.

Strathclyde Regional Council (1994), *Environmental Action in Strathclyde. Charter for the Environment Strathclyde,* Chief Executive Department.

The Herald (1997), 'Scottish Office and Cosla in new accord', (22. August).

The Economist (1998), 'Train, planes, but not automobiles', (6. June).

The Scotsman (2001), 'Scotland could go it alone by 2013' and 'Labour gets new poll warning', website.

Wegner, H.-A. (1993), 'Die Umweltpolitik der EG im Spannungsfeld zwischen Harmonisierungszwang und Subsidiaritätsprinzip', *Berichte der Bayerischen Akademie für Naturschutz und Landschaftspflege* Nr. 17 (Sonderdruck).

6 Case Study: Implementing the EIA Directive in Scotland and Bavaria

Introduction

The previous Chapters investigated the three government *layers* involved in the *filtering* process of EU environmental policies. The Chapters highlighted key problems which contribute towards the EU environmental policy implementation deficit. The following case study examines the EU environmental policy 'reality' further by focusing on the *filtering* process of one particular piece of legislation: the Council Directive of 27 June 1985 on the assessment of the effects of certain public and private projects on the environment (85/337/EEC) (in short environmental impact assessment or EIA Directive).[1] The actual field research for this case study was conducted in 1997. Since then two significant changes have occurred which have to be taken into consideration:

- the EU adopted Directive (97/11/EC) amending the EIA Directive
- Scottish devolution has had legal and practical implications for the EIA policy in Scotland.

At this preliminary stage (the deadline for the 1997 Directive's formal transposition was in 1999), it appears that a few adjustments have been made following the above changes of the late 1990s. Indeed, Member States and their *sub*national regions have (to a greater or lesser extent) adjusted their legal and practical frameworks to accommodate the new EIA requirements. While these adjustments have changed the picture slightly, by and large the key characteristics and obstacles in the *filtering* of EU environmental policies are still evident. The following case study therefore takes into consideration the above changes but will concentrate on the original EIA Directive and its formal and practical implementation.

The EIA Directive of 1985 was selected for essentially two

reasons. Firstly, the EIA Directive featured characteristics common to other EU environmental policies such as the Habitats Directive and water quality Directives. For instance, it included a 3-year deadline by which Member States were required to adjust their standards to achieve the environmental objective of 'minimum-regret-planning'. As is typical for EU environmental Directives, the EIA requirements were not met by the deadline by most Member States for political, legal-administrative and economic reasons.

Having stressed some of the similarities with other Directives, the EIA Directive stood out in other areas and has been described by many as one of the most complex and controversial EU environmental policies to date.[2] With the EIA Directive, (then) EC policy-makers entered new territory by setting environmental standards in the planning policy area, an area that had previously been the exclusive domain of national and *sub*national actors. The EIA Directive affected a variety of planning projects which fell within the competency of *sub*national (and local) government. Not only did the EIA Directive challenge existing national and *sub*national planning policies and competencies, it also tended to clash with economically motivated projects such as road construction, housing developments and the construction and operation of manufacturing plants. All these factors rendered the Directive's implementation difficult.

Apart from uncovering unique and typical implementation characteristics, the EIA case study also promised to highlight the government *layers* involved as well as their 'inter-connectedness' during the process. Furthermore, it promised to highlight the *layers'* divergent *determinants* and their influence on the success (or failure) of an EU environmental policy.

The EIA Directive: Origin and Objective

The EIA Directive has been one of the most discussed EU environmental policies. Over the years, many analysts have highlighted the complexity of the policy, discussed its legal and practical implications and described the subsequent difficulties with which practitioners processed the policy.[3] Indeed, the EIA Directive has been controversial and a tough nut to crack.

The adoption of the EIA Directive in 1985 followed a long and cumbersome negotiation process. Initially, all actors agreed that a common EIA policy was paramount for both environmental considerations (i.e. restricting environmentally damaging project developments) and economic considerations (i.e. level-playing-field as opposed to 'environmental dumping' in the planning policy area). At the same time, concerns were raised about a further loss of discretionary powers, the potential administrative and financial costs associated with environmental assessments, and the prospect of more 'red tape' on business developments. The Danish and UK Governments in particular, found it hard to accept a European policy which interfered with traditionally national and *sub*national decision-making. The adoption of the EIA Directive was therefore complex (see Chapter 3) involving various actors and interests from different government levels in a complicated bargaining process. Consequently, the end product of the intense bargaining was a Directive which represented a more 'digestible' compromise of considerations.[4]

The Directive of 1985 consisted roughly of four procedural stages:

- Firstly, projects which were potentially harmful for the environment and therefore required an EIA had to be identified. Under Annex I, the Directive provided a list of projects which had 'significant effects on the environment' and which required 'as a rule' a systematic assessment. The Directive also contained an Annex II which referred to projects which 'may not have significant effects on the environment in every case' but 'should be assessed where the Member States consider that their characteristics so require'.

- Secondly, once it was established that an EIA was necessary for a project application, the developer seeking planning permission was required to provide relevant information on the project in the form of environmental statements. Annex III of the Directive outlined a detailed list of environmental information items which developers were required to produce.

- Thirdly, having provided all the necessary information, interested parties and the public had the opportunity to participate in the planning process. According to the Directive, Member States were to ensure that any request for development consent and any information were made available to the public, and that the public concerned were given the opportunity to express an opinion before the project was initiated.

- And finally, a planning decision could only follow *after* the three stages of identification, environmental statement and information/ consultation had been accomplished.

The EIA Directive's main purpose was to oblige Member States' planning authorities to consider environmental aspects in planning procedures as early as possible whenever a proposed project was likely to have a major effect on the environment. The idea behind environmental impact assessment was to conduct 'minimum-regret-planning' which involved the identification, evaluation and incorporation of environmental externalities during the planning process so that 'unreasonable' environmental damage could be avoided. The Directive itself did not provide for specific environmental criteria; it did not require Member States and their planning authorities to change their policy priorities and 'become green'. The Directive required planning authorities to adjust their formal procedures and thereby integrate environmental interests more strongly into planning considerations so that 'environmentally sensible' decisions were taken. Legally, the Directive established a right for the public to be consulted before permission was given to projects which may have had damaging implications for the environment. In practice, however, political-administrative actors and the judiciary were unable (and often reluctant) to bind developers and planning authorities into a coherent EIA system. The EIA Directive left much of the ways and means to achieve 'minimum-regret-planning' to the Member States and their implementors. It provided considerable scope for interpretation, particularly in areas such as time limits and methods of consultation. Consequently, the effectiveness of the policy depended upon the national and *sub*national implementors' ability and willingness to implement and enforce the policy within their political-administrative systems.

The EIA Directive has since been amended. Directive (97/11/EC)[5] intends to bind Member States and their *sub*national regions more thoroughly into a harmonised framework. To make sure that potentially harmful projects do not fall through the EIA net, the new Directive adds more project categories to Annex I and Annex II, while more specific criteria (in Annex III) are provided for the identification of Annex II projects that require EIA. The Directive obliges planning authorities to 'screen' project applications 'case-by-case', i.e. not automatically exclude projects if they fall below certain thresholds. The Directive also includes a provision which requires planning authorities to outline the 'scope' of EIAs for developers. The latter provision, however, is not mandatory and only sets in if developers request so. Finally, the Directive requires planning authorities to explain why they initiate, or decide not to initiate, an EIA.

The Commission had intended a more ambitious EIA amendment,

however, similar to the 1985 Directive, a moderate compromise was adopted in 1997 by the Council (and the EP), which allows for loopholes, notably in the area of 'scoping'. The Directive's deadline for transposition was in March 1999 and, so far, not all formal and practical obligations have been met by many Member States. The reasons for the delay and difficulties are similar to those that prevented the 1985 Directive from being implemented properly (outlined below).

As for any EU environmental Directive, the EIA Directive of 1985 required formal transposition, practical implementation and (if necessary) legal enforcement within the Member States and their *sub*national regions. Accordingly, the following Sections investigate the three steps of implementation in the national and *sub*national *layers* and assess how and to what extent implementors in Scotland and Bavaria have processed and influenced the EIA Directive. The Sections compare implementation performances in Scotland and Bavaria and contrast their performances with experiences in the UK and Germany at large.

The UK Layer and the EIA Directive

Considering that the UK Government initially opposed the adoption of Directive (85/337/EEC), the first stage of the implementation process, i.e. the formal transposition of the Directive, was conducted relatively swiftly and problem-free within the national *layer* (for favourable *determinants* see figure 6.1). UK policy-makers concluded that the EIA Directive did not require parliamentary scrutiny or formal approval. In addition, the policy's integrative approach to consider a wide range of inter-connected environmental impacts during planning fitted-in well with the UK's broad definition of the environment. The Directive text was therefore adopted promptly and verbatim by the (then) two central government ministries, the DoE and the Scottish Office, into existing planning policy frameworks in the form of Statutory Instruments.[6]

Although part of a centralised state system at the time, Scotland required separate transposition from the rest of the UK for several reasons. Scotland's legal system differed (and continues to do so) in many respects from the system in England and Wales. In addition, policy matters such as environmental policies were dealt with by the territorial Scottish Office and the Secretary of State for Scotland. Scotland also differed from the rest of the UK in terms of natural habitats, infrastructure, population density and industrial output. Its environment was (and still is) particularly affected by sectors such as salmon farming and wind generation, which cannot be found in England and Wales in such a large scale. While all these aspects necessitated separate formal transposition of the EIA Directive for Scotland, Scottish Office officials did not obstruct (or divert from) the first stage of the *filtering* process. In fact, the Scottish Office followed the DoE lead and adopted an EIA Statutory Instrument (SI) which differed only marginally from the England/ Wales documents.[7] While differences were only marginal, the Scottish Office nevertheless translated the EIA policy into the *sub*national (i.e. Scottish) context and gave the policy its 'personal touch' (the details of the Scottish Office SI and Circular are outlined below). By doing so, the Scottish Office followed its 'hybrid' function as a central government ministry representing both UK and Scottish interests. In this sense, both national and *sub*national *layers* overlapped.

The formal transposition of the follow-up Directive (97/11/EC) coincided with Scottish devolution which, however, did not hinder or delay the transposition process. While taking over from the 'hybrid' Scottish Office, the Scottish Executive did not divert from rest and transposed the Directive in time in the form of the Environmental Impact Assessment (Scotland) Regulations 1999 accompanied by a Circular and a Planning Advice Note. England and Wales transposed the Directive with the Town and Country Planning (Environmental Impact Assessment) (England and Wales) Regulations 1999 and for England the DETR published another Circular.[8] Even though the immediate overlap between UK and Scottish *layers* disappeared, business remained very much as usual as far as the formal transposition of EU environmental legislation was concerned.

While the formal transposition of the 1985 Directive was relatively swift, practical implementation of the policy in the UK proved to be more problematic. Enjoying almost limitless discretionary room in determining EIA cases (a fact that contributed to the Directive's amendment in 1997), planning applications were processed on an informal case-by-case basis. Furthermore, a number of UK-wide EIA policy studies[9] revealed insufficiencies in the quality of EIAs (e.g. many vital environmental criteria were not mentioned in environmental statements) as well as quantity of EIAs (e.g. many potentially harmful projects were not identified). Certain formal and informal *determinants* represented major obstacles for the EIA policy in the UK: the above studies identified as the main problems in EIA policy implementation insufficient expertise, lack of resources, limited interest and lack of commitment. More recent studies showed that the EIA practice improved marginally since 1988.[10] As one of the key reasons, researchers identified a (slow) learning process in dealing with EIA. While initial EIA policy results were disappointing, the DoE was adamant to provide detailed guidance for planners and developers to improve EIA standards and thereby demonstrated a supportive and positive attitude towards the EIA policy. By and large, this supportive attitude was missing from the Scottish Office (see Section below).[11]

In terms of legal enforcement, UK judges were generally uneasy about the EIA policy. More specifically, they were reluctant to support the policy with additional criteria and guidelines where the Directive and its national legislation were 'silent'. For England and Wales, Ward and Alder observed that courts did not interpret the EIA Directive in a sympathetic light, i.e. they did not take account of the wider concept and idea of the Directive, nor did they establish a direct effect.[12] It was therefore up to the planning authorities to pursue and implement the objective of 'minimum-regret-planning' within their local communities. Directive (97/11/EC) is intended to close these interpretative gaps and responds to the courts' reluctance to deal with 'silent' aspects of EU legislation.

In sum, the UK national *layer* featured some favourable formal and informal conditions (or *determinants* outlined in figure 6.1) for the *filtering* of the EIA Directive. The (then) centralised political-administrative system as well as the legal framework of the UK accommodated the integrative environmental policy without major difficulties. In addition, the (then) DoE played a pro-active role in the implementation of the policy. However, the policy faced other formal and informal *determinants* which hindered its the successful implementation into the UK planning policy practice. These *determinants* concerned mainly problems of commitment and resources on the part of planning authorities. In addition, the UK legal system was unable (and arguably unwilling) to back up the EIA policy; the judiciary did not fill legal gaps were the Directive and Statutory Instruments remained silent. In the meantime, some of these gaps have been filled by Directive (97/11/EC).

Figure 6.1: The UK Layer and the EIA Directive

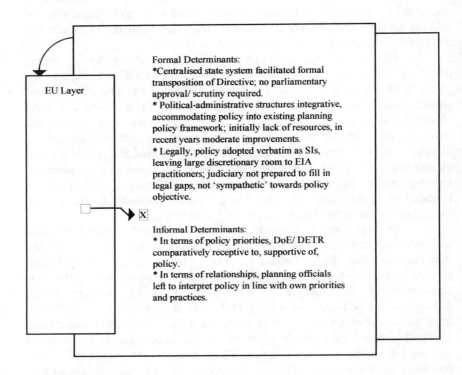

EU Layer

Formal Determinants:
*Centralised state system facilitated formal transposition of Directive; no parliamentary approval/ scrutiny required.
* Political-administrative structures integrative, accommodating policy into existing planning policy framework; initially lack of resources, in recent years moderate improvements.
* Legally, policy adopted verbatim as SIs, leaving large discretionary room to EIA practitioners; judiciary not prepared to fill in legal gaps, not 'sympathetic' towards policy objective.

X

Informal Determinants:
* In terms of policy priorities, DoE/ DETR comparatively receptive to, supportive of, policy.
* In terms of relationships, planning officials left to interpret policy in line with own priorities and practices.

The Subnational Layer: Scotland and the EIA Directive

Formal Transposition

As mentioned above, the (then) Scottish Office followed the DoE lead and did not divert substantially from the rest of the UK with its formal transposition of Directive (85/337/EEC). However, closer examination of the Scottish Office Circular (13/1988) (a document which accompanied, explained and summarised the Scottish SI) reveals that the Scottish Office perceived and processed the policy in a manner distinct from the rest of the UK. In essence, the Scottish Office was reluctant to adjust Scottish planning practices and restrict economic development in Scottish local communities for the sake of an EU policy which promoted an environmental level-playing-field in planning (see informal *determinants* in figure 6.2).

According to the Circular, the development control system already in place in Scotland covered the main objectives of the Directive and only minor additions were considered necessary to implement the Directive fully.[13] Instead, the Scottish Office placed great emphasis on the date of applicability of the EIA policy (i.e. project applications in progress at the time of the Directive's formal transposition were excluded) and the developers' means of appeal against EIAs. The Scottish Office Circular elaborated in great detail the question whether a planning application required an EIA, indicating that the impact of the policy on costs and administration should be kept to a minimum. Once a planning authority in Scotland was informed by a developer about a forthcoming planning application, the planning authority had four weeks time to investigate whether the proposal in question fell within the project categories of Annex I and II. If the proposed project was listed under Annex I, an EIA was mandatory. If the project could be found in the Annex II category, the planning authority had to decide whether the project was likely to have significant effects on the environment and therefore required an EIA. The advice given on Annex II criteria was only indicative.

An EIA was required if:

- the project was of more than local importance
- the project was situated in a 'particularly sensitive or vulnerable location'
- the project was 'unusually complex' and had 'potentially adverse environmental effects'.

The indicative thresholds suggested by the Scottish Office in Annex C of the Circular did not provide further guidance for planning authorities. As a result, Scottish planning authorities were free to consider Annex II projects on a case-by-case basis and could take decisions in accordance with their (economic) policy priorities and strategies.

Once the planning authority came to the conclusion that an EIA was required, it had to provide reasons for its decision. The planning authority was not obliged to give reasons if it was of the opinion that a project did not significantly affect the environment and did not require an EIA. Developers were therefore in an advantage over environmentalists. They had the opportunity to find out why planning authorities were asking for environmental statements and could change their application strategy accordingly. Opponents of project applications did not enjoy the same early access to information which could have strengthened their position in the process.

The Scottish Office Circular elaborated on the developer's right of appeal against an EIA decision which enabled him to refer his case to the Secretary of State for Scotland for direction. The Secretary of State could either confirm the planning authority's opinion that an EIA was necessary, in which case an environmental statement was obligatory. He could also disagree with the planning authority, in which case the developer was not obliged to produce a statement. Therefore, as a central government minister (representing the national *layer*), the Secretary of State played a significant role in the planning process in Scotland: he could over-rule any decision taken by a Scottish local authority.

Environmental statements were covered in only one comparatively small section of the Circular which provided much discretionary room for developers. The briefness of the section suggests that the financial and administrative burden for both developers and planning authorities were to be kept at a tolerable level. The Circular's guidelines did not mention all the requirements listed in Annex III of the EIA Directive. They did not include an outline of the main alternatives considered by the developer, neither did the guidelines include a description of the developer's research methods and process.[14] Instead, the Circular emphasised that:

> [t]here [was] no statutory provision as to the form of an environmental statement.

In other words, planning authorities were not in the position to make any judgements on the form and quality of developers' environmental statements; they could only 'request' more information if a statement did not provide sufficient information for an EIA.

Following the submission of the environmental statement by the developer, the planning authority was required to inform the public and statutory consultees about the planning application. The planning application together with the environmental statement were advertised in the local press and the *Edinburgh Gazette*. The notices had to indicate where and when the environmental statement could be inspected by the public. Statutory consultees received copies of environmental statements free of charge, while other interested parties could request paying a 'reasonable charge' to cover the costs of production. The list of statutory consultees was restricted to 'relevant bodies' such as adjoining planning authorities, Scottish Natural Heritage (SNH), and the Secretary of State for Scotland who was to be informed whenever proposals were likely to affect water supplies, waste disposal, noise and air pollution, trunk roads and special roads, historic buildings, and Royal Parks and Palaces.[15] Statutory consultees and other interested parties had the opportunity to comment on planning applications by submitting written representations about the proposed development within four weeks.

The Circular stated that planning authority should inform the developer which bodies had been consulted and should send copies of the planning application and the environmental statement to the Secretary of State for Scotland. Again, the discrepancy between applications with and without environmental statement was striking: planning authorities were required to send only applications which had environmental statements attached, they were not required to forward applications where environmental statements were considered as unnecessary. Many potentially harmful projects could therefore fall through the net of EIA scrutiny.

Having accomplished the formalities of the preceding three stages, authorities could attend to the actual planning decision. Taking into account the environmental statement and any comments from consultees and interested parties, planning authorities had to take a decision within 16 weeks, according to the Circular. If the information provided was not adequate, planning authorities could request further information within the 16 weeks time limit. Once a decision was taken, planning authorities were required to notify developers, the Secretary of State for Scotland, and statutory consultees. The Circular did not mention in which form the planning authorities should notify interested parties of their decisions, i.e.

whether or not reasons for the decision were to be included. Further, only appeals to the Secretary of State for Scotland by developers were considered at a 'post decision' stage; possible appeals put forward by the public were not mentioned in the Circular. And, again, the Secretary of State could over-rule an authority's decision if he disagreed with the planning authority.

Although the policy's preventive objective was mentioned in the introduction of the Scottish Office Circular, the concept of 'minimum-regret-planning' was not evident in the remaining parts of the document. Throughout, the authors were almost apologetic about additional work and costs resulting from the EIA policy. They emphasised that 'no unnecessary burden' should occur for planning authorities and developers and that:

> additional costs imposed on developers by the requirement to provide information about environmental effects should be kept to a reasonable minimum.

The text lacked any similar sensitivity towards advocates of environmental concerns and their problems in representing 'green' interests.

In 1994, the Scottish Office had to up-date the list of Annex II projects as a response to amendments made by the DoE for England and Wales. Three new project categories (wind generators, motorway service areas, coastal protection works) were adopted with considerable reluctance. In Circular (26\1994) the Scottish Office stressed that other categories (salmon farming, water treatment plants, non-motorway service areas and golf courses) were discussed but not included. A combination of considerations explain the Scottish Office's behaviour: the Scottish Office followed both the national *layer's* (i.e. the then Conservative Government's) policy of deregulation *and* the economic interests of certain private and public sectors in Scotland. In this case, the representation of national and *sub*national interests by the Scottish Office was not paradoxical (see Chapter 5), but merged into a convenient combination of economic interests in both *layers*. The adoption of three new categories was the only concession the Scottish Office was willing to accommodate environmental considerations. Although the other four categories were not included in the Annex II list, the Circular nevertheless advised planning authorities to consider 'voluntarily' their environmental impacts.

In comparison with the DoE, the Scottish Office took a minimalist and defensive approach towards the EIA policy. Apart from an apologetic Circular, the Scottish Office produced a two-page leaflet which did not match up with the DoE documents which guided planners and developers in the rest of the UK.[16] Figure 6.2 describes the formal *determinants* as favourable to the transposition of the EIA Directive in Scotland. Although Scotland transposed the policy separately from the rest of the UK, the (then) centralised state system and legal framework allowed Scottish implementors to transpose the EIA Directive verbatim without parliamentary scrutiny and policy amendments. On the other hand, unfavourable informal *determinants* were evident right from the beginning of the Directive's transposition into the Scottish context. Informal obstacles occurred mainly in the form of economic considerations, in particular the financial costs and fears over more 'red tape' for Scottish business developments. These considerations compelled Scottish Office officials to keep additional and supportive measures for the EIA Directive to a minimum and apologise for any 'inconvenience' caused by the policy.

Devolution did not pose formal obstacles for the transposition of the 1997 Directive. In fact, the new Directive was integrated into the planning policy framework without delay. Obviously, procedures were adjusted so that Scottish Executive Ministers (i.e. Ministers responsible for rural affairs, transport etc., depending on policy area affected) would replace the Secretary of State for Scotland in his capacity of dealing with Scotland-wide and problematic ('appeal') EIA cases. Interestingly, the new Directive was considered by the Scottish Executive predominantly as a means to avoid some of the inconveniences of the 1985 Directive. In contrast to their Bavarian colleagues at the StMLU, who complained bitterly about further 'unnecessary' administrative obligations and 'red tape' from the 1997 Directive, Scottish Executive officials highlighted the new criteria, or thresholds, which would help determine quickly whether an Annex II project required EIA.[17] For the Scottish Executive this element of the Directive facilitated a move away from a case-by-case identification process to a more threshold-based process. Whether or not this interpretation will result in some (potentially harmful but below-threshold) Annex II projects being excluded from screening remains to be seen. On paper, the Scottish Executive has accommodated the new Directive and adjusted Scottish identification and consultation procedures accordingly.

Practical Implementation

At first glance, the implementation of the 1985 Directive in Scotland did not diverge significantly from the implementation in the rest of the UK. But a closer look at the details of practical implementation reveals key differences in the Scottish EIA practice. Smith (1990),[18] for instance, pointed out that Scottish planners and developers already had considerable experience in the environmental assessment and management of North Sea oil projects, but also noted that this advantage was not utilised for other planning sectors.[19] More importantly, Smith observed that Scottish Office officials made no secret of their criticism over the costs and additional work resulting from the EIA Directive's requirements. Many Scottish practitioners (i.e. planners and developers) would later confirm this attitude by applying only a bare minimum of EIA requirements.

The author's own research findings reflect the Scottish Office's open dislike of the EIA policy as one of the policy's main informal obstacles. In fact, considering the initial response to research enquiries, the EIA Directive did not appear to have a good start in Scotland: one Scottish Office key official described the Directive as 'an awful thing' but also admitted that he had 'never read that thing'.[20] Nevertheless, local planning authorities received brief EIA policy guidance from the Scottish Office.[21] In return, local planning authorities provided the Scottish Office with a list of planning procedures requiring EIA and thereby generated a rough overview of the Scottish EIA practice.

Scottish planning authorities viewed the EIA policy with mixed feelings. Some planning authority officials complained about the increased bureaucracy and high expectations on the part of environmentalists, some were indifferent about the policy, while others approved of their 'new' strengthened authority to request environmental information from developers. Although planning authorities consulted the Scottish Office Circular on a regular basis, implementation practice depended very much upon informal *determinants*, in particular planning authorities' attitudes and preferences with individual planning officers leading the EIA process.[22] As a result, one of the policy's aims, namely the harmonisation of environmental standards in planning, was not achieved *inside* the Scottish *layer*.

The SI and the Circular left Scottish planning authorities with considerable discretionary room concerning the 'screening' and identification of potentially harmful projects. Consequently, many planning officers tended to avoid obstacles to economic or other developments in their communities which they perceived as inconvenient and unnecessary. The number of EIAs was therefore limited to project applications which were 'obviously' harmful. Taking rough estimates of the project's size and location, planning officers checked whether applications belonged to either

Annex I or Annex II. Potentially harmful and controversial projects such as quarries, incinerators, or waste disposal plants were considered for EIA, while other, less obvious but equally harmful, projects were not checked at all. Environmental interest groups and other consultees who could have helped identify harmful projects, were not consulted by planning authorities during the screening process. Enjoying considerable discretionary powers, planning officers were reluctant to hinder economically lucrative developments in their local areas. This situation was worsened by the fact that planners showed a lack of experience and resources[23] in identifying potentially harmful projects. Directive (97/11/EC) may bring planning authorities more into line with regards to the identification of Annex II projects requiring EIA. The actual impact of the new Directive on practical implementation is difficult to assess at this stage since formal transposition was completed only in 1999. However, first indications are that the new Directive has not changed planning practices significantly (see Scottish Executive figures below).

Returning to the 1985 Directive, there were differences in opinion concerning the information provided by Scottish developers. Generally, Scottish planning authorities were satisfied with the content and quality of environmental statements, although one planning officer admitted that statements were biased and focused on information in favour of the projects.[24] In contrast, environmental interest groups and some observers considered the quality of environmental statements as generally poor.[25] The most disappointing results were found in the Annex III sections 'alternatives' and 'remedial solutions', aspects which hardly received any consideration. Aware of the statements' insufficiencies, planning authorities nevertheless avoided additional work and only rarely returned applications to developers with the request for more information.[26] Therefore, (self-) interests in minimising the work-load prevented planning officials from pursuing a more rigorous approach towards environmental statements.

Scottish planning authorities made full use of their discretionary powers in determining the list of consultees. Consequently, potential opponents of project applications had to rely upon the *Edinburgh Gazette* and local papers to find out about planning applications. Statutory consultees and 'affected' parties were informed properly, according to Scottish planning authorities. Interestingly, the majority of consultees contacted by the authorities did not represent environmental interests as such but interests of traditional lobby groups, local communities and economic sectors.[27] Planning authorities were generally reluctant to open the EIA process to 'outsiders'. Among other reasons, planning officers feared the increased work-load caused by a wider audience. At the same time, planning authorities were also aware that environmental interest groups could jeopardise projects through increased public and media pressure. As a response, environmental interest groups were allowed to participate, but only if they explicitly requested so (see 'relationships between actors' in figure 6.2). Given that the 1997 Directive is not covering the choice of consultees (only their right to be informed of planning authorities' reasons for their decisions), any changes in the planning authorities' consultation practices are highly unlikely.

In the years that followed the formal transposition of the 1985 Directive, Scottish planning authorities rejected a number of planning applications. However, it is difficult to establish exactly what impact the EIA policy had on these planning decisions. Again, some planning officers stated that the introduction of the EIA policy had a positive influence on planning decision-making. The policy provided planning officers with more authority to demand environmental information and highlight environmental dangers of a project. In some cases, the information generated by EIA *contributed* towards either the projects' modification or even the refusal of planning applications.[28] Other planning officers, however, noticed no difference in decision-making following the introduction of the policy.[29] In some cases, planning authorities openly supported planning applications for economic reasons, with the result that EIA findings were ignored completely during decision-making.[30] Overall, there was no coherent (level-playing-field) pattern in the application of the EIA policy in Scotland. The policy depended very much upon individual planners and their priorities and attitudes towards the EIA policy objective of 'minimum-regret-planning'. This picture has been confirmed by Scottish Executive figures of 1999, which show that despite the (then impending) harmonisation measures of the 1997 Directive some planning authorities conducted a comparatively large number of EIAs (e.g. between April 1998 and March 1999 Highlands Council conducted 43 EIAs), while others kept their EIAs to a minimum (e.g. both East Dunbartonshire and East Renfrewshire had no EIA in the same period).[31]

Enforcement

In terms of EIA policy enforcement, court decisions in Scotland did not differ significantly from decisions in England and Wales. Figure 6.2 indicates that the Scottish legal system, too, was 'unsympathetic' towards the EIA policy. In one particular instance, a Scottish judge stressed that the EIA Directive and its Scottish Statutory Instrument did not provide clear-cut provisions to determine the question whether a project listed in Annex II required EIA.[32] Planning authorities in Scotland (and indeed the UK in general) traditionally enjoyed independence in local planning. Since vital elements of the EIA Directive (such as the question whether an Annex II project required EIA) were left to the implementors' discretion, it was difficult, if not impossible, for project opponents to legally challenge planning decisions on the grounds that an EIA was inadequate. It remains to be seen whether the provisions of the new Directive are 'sufficiently clear-cut' for Scottish courts to interfere with planning authorities' decisions.

Overall, the EIA Directive was *filtered* through the *layers*; however, most of its practical implementation took place *within* the Scottish *layer*. True, the Directive's formal transposition was conducted by (then) central government ministries: the Scottish Office and the DoE. While Scottish Office officials followed the national *layer's* policy line, they also translated the policy into the Scottish (i.e. *sub*national) context. In this sense, national and *sub*national *layers* overlapped. The Scottish Office made no secret of its reluctance to adjust planning processes and shift (economic) priorities. Accordingly, the EIA Directive altered Scotland's planning practices only moderately and depended upon individual planning officers' attitudes towards environmental considerations and other interests such as housing and economic development. Some planning officers used the EU policy as a means to take 'environmentally-friendly' decisions.[33] Other planning officers took a more critical view of the policy and were able to minimise the policy's impact on planning decisions.[34] In either case, a level-playing-field of minimum-regret-planning was not evident inside the Scottish *layer*.

A level-playing-field is still not evident in Scotland, despite the new Directive's emphasis on Annex II harmonisation. It is perhaps too early to assess the implications of the 1997 Directive on planning practices in Scotland.[35] What can be said at this stage is that the formal transposition of the Directive and its integration into the Scottish planning framework was conducted swiftly and that only a few minor alterations were deemed necessary.

Figure 6.2: The Scottish Layer and the EIA Directive

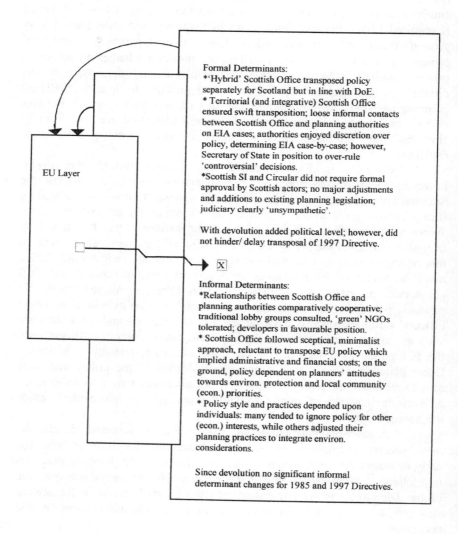

Formal Determinants:
*'Hybrid' Scottish Office transposed policy separately for Scotland but in line with DoE.
* Territorial (and integrative) Scottish Office ensured swift transposition; loose informal contacts between Scottish Office and planning authorities on EIA cases; authorities enjoyed discretion over policy, determining EIA case-by-case; however, Secretary of State in position to over-rule 'controversial' decisions.
*Scottish SI and Circular did not require formal approval by Scottish actors; no major adjustments and additions to existing planning legislation; judiciary clearly 'unsympathetic'.

With devolution added political level; however, did not hinder/ delay transposal of 1997 Directive.

Informal Determinants:
*Relationships between Scottish Office and planning authorities comparatively cooperative; traditional lobby groups consulted, 'green' NGOs tolerated; developers in favourable position.
* Scottish Office followed sceptical, minimalist approach, reluctant to transpose EU policy which implied administrative and financial costs; on the ground, policy dependent on planners' attitudes towards environ. protection and local community (econ.) priorities.
* Policy style and practices depended upon individuals: many tended to ignore policy for other (econ.) interests, while others adjusted their planning practices to integrate environ. considerations.

Since devolution no significant informal determinant changes for 1985 and 1997 Directives.

EU Layer

The German Layer and the EIA Directive

In comparison with the UK and Scotland, the Federal German constitutional setting, political-administrative structures and legal system rendered the formal transposition of the EIA Directive extremely difficult (for formal *determinants* see figure 6.3). In fact, the transposition turned out to be a legal nightmare involving long deliberations over the form and content of both Federal and Länder legislation. Among other issues, discussions surrounded the question whether the EIA Directive deserved a separate piece of legislation or whether the EIA policy should be integrated into the existing legal framework. Following a heated debate (the Bundesrat forwarded 59 amendments to the EIA Federal Law proposal of which only 30 were accepted by the Bundestag), the EIA Directive was formally transposed two years after its deadline. The national *layer* (i.e. the Bundestag and Bundesrat) adopted a Federal EIA law, 'Gesetz über die Umweltverträglichkeitsprüfung' (UVPG),[36] and amended sixteen existing Federal Laws affected by the new requirements.[37] Some amendments of these sectoral laws only came into force in 1992 (amendment to the Federal Emission Law) and 1994 (amendment to the Nuclear Safety Law). Apart from the UVPG and amendments to existing legislation, Federal legislators felt that the EIA policy further required detailed and complementary guidelines in the form of 'Verwaltungsvorschriften' (administrative regulations) which were eventually formalised in 1995.[38]

Two years after the deadline, German legislators are still in the process of transposing the 1997 Directive. Parts of the Directive were intended for the 'Umweltgesetzbuch' (a comprehensive environmental law book currently in preparation) while other parts were intended for a separate amendment law to the UVPG. Some project categories were to be transferred from UVPG to Umweltgesetzbuch while others were to remain in the original text. Until the formal transposition process is completed, further guidance on the new EIA policy is provided with 'guidelines' from the BMU issued in June 1999. For Commission officials these guidelines are obviously not sufficient compelling them to put more pressure on German legislators to transpose the EIA policy (i.e. both 1985 and 1997 Directives) quickly and properly. In response, Federal Minister for the Environment Trittin proposed recently to speed up the process by ignoring the Umweltgesetzbuch for the time being and instead integrate the policy into the existing legal framework.[39] While these considerations continue, the case illustrates yet again that the transposition of EU environmental policies in Germany is rather thorough but also piecemeal, cumbersome and slow.

Focusing on the EIA procedural steps, the first stage of the 1985 Directive posed major difficulties for Federal legislators. While the project list of Annex I was adopted verbatim from the EU Directive text, the Annex II list proved incompatible with German legal-administrative tradition. Federal legislators found it difficult to transpose a measure which provided flexibility and discretionary judgement instead of water-tight rules and regulations. It was therefore decided to include separate paragraphs regulating matters such as airports, nuclear safety and mining in the UVPG, and transfer other Annex I and some Annex II items to one comprehensive and detailed list of projects which should, as a rule, undergo EIAs.[40] In effect, these departures from the EU Directive caused not only a delay in the implementation of the EIA policy for certain categories, but also excluded projects from the Federal legislation list which, according to the EIA Directive's Annex II, should have been considered by planning authorities as potentially harmful.[41] Over the years, the Commission has tried repeatedly to force German legislators to include these categories, however without much success.[42] The situation is likely to worsen with the new Directive, which not only superimposes a new list of Annex II thresholds but also requires planning authorities to adopt, in conjunction with these thresholds, a 'case-by-case' approach in the 'screening' process.

Where German legislators did lead the way in the EIA policy was in the area of 'scoping'. In line with the German legal tradition to 'regulate' environmental standards, much attention had been drawn to the formulation of 'scoping' standards for the 1985 Directive. UVPG Paragraph 5 and supplementary administrative regulations elaborated on scoping and the content of environmental statements. If applied correctly, the regulations did not provide much room for manoeuvre for developers. On the other hand, some of the environmental statement items were required only if their inclusion was 'zumutbar' ('reasonable') for developers. Requested information on environmental surroundings, project alternatives and research difficulties should therefore not exceed 'unreasonable' quantitative and qualitative thresholds. The question remained (and presumably will remain in future) where to draw the line between reasonable and unreasonable information. It was up to administrative courts to determine the adequate quantity and quality of environmental information in each case.

Federal information and consultation procedures remained unchanged. According to Paragraph 9 of the UVPG, public consultation should be conducted by following the procedures of existing legislation, in particular Paragraph 73 of the Law on Administrative Procedures (Verwaltungsverfahrensgesetz or VerwverfG). The latter stated that planning authorities should inform the public one week prior the actual consultation process that planning application documents were available for inspection. The documents were to be displayed for the period of one month in the local communities affected by the projects, after which the public had two weeks to comment. The subsequent consultation process excluded the general public: the planning authority invited to a consultation meeting only parties directly affected by the project, i.e. individuals who had a legal or material interest in a project application. Therefore, challenges against a project could only be made on the grounds of material damage (for instance, a motorway running through private property) or infringement of legal rights. Other project opponents such as environmental interest groups could advise 'affected' individuals but had no formal right to participate in the consultation process. Only in 'high risk' areas such as nuclear energy, the Federal legislator considered 'jedermann' ('anybody') affected and eligible to participate at every stage of the consultation process. This restricted access for the public could result in the neglect of certain environmental aspects, but was in accordance with the EIA Directive which left the details of consultation to the Member States.

Once the consultation process was completed, the planning authority was required to summarise the evidence within one month. The authority then assessed the application together with the information provided by various parties. The assessment was to be conducted strictly in accordance with quantitative thresholds established in Federal (and Länder) Laws and in line with the detailed guidelines of administrative regulations which supplemented the UVPG. Aspects that fell outside threshold criteria, such as cross-media and accumulative impacts, were not specifically mentioned in Federal (and Länder) legislation. Once a planning authority decided to permit a project, it was required to include the reasons in favour of the project. In the case of planning refusal, the decision itself was sufficient for information. Similar to the consultation process, only 'affected parties' who commented on the project application were informed about the planning authority's decision.

Federal legislators accepted the objective of the EIA policy and specifically referred to it in the first Paragraph of the UVPG. However, the EU Directive's integrative elements rendered the policy's *filtering* process almost impossible in a legal-administrative system which was (and still is) compartmentalised and sector-orientated. The Umweltgesetzbuch, which is currently being prepared, is intended to tackle this fragmentation and introduce a more integrated policy approach. However, indications are that the new environmental law book will not eliminate this problem completely.

From 1985 onwards, legislators in both national and *sub*national *layers* conducted a difficult transposition whereby the EIA Directive was scrutinised by parliament and shaped to fit Federal and Länder legal-administrative preferences. However, unfavourable formal *determinants* were not the only obstacles during the *filtering* of the policy. Weber and Hellmann commented that the formal transposition of the 1985 Directive in Germany could have been 'gemeinschaftsfreundlicher' ('more euro-friendly').[43] This assessment indicates that Federal and Länder legislators also lacked the informal political commitment to adjust their planning systems to accommodate the EIA Directive (Bavarian legislators' lack of commitment is described below). Considering the difficulties with which legislators transpose the 1997 Directive at the moment, formal and informal *determinants* have not changed significantly in recent years.

As far as the practical implementation of the 1985 Directive is concerned, the picture looks equally fragmented, legalistic and technocratic. For these reasons, it has been difficult to gain a comprehensive overview of the policy's practice in Germany. From the information available, it is evident that the UVPG project list excluded certain categories mentioned in the Directive. In addition, the information provided by developers was often too detailed and difficult to understand (a situation made worse by the fact that Federal legislation did not require developers to produce a non-technical summary). Further, the consultation process was restricted to certain groups with material interests and planning decisions were predominantly based on quantitative thresholds while many cross-media aspects of pollution and environmental deterioration were neglected by Federal practitioners.[44]

Since the mid-1990s, German legal experts have taken stock of the EIA policy and its prospects of enforcement. One conclusion was that neither the high expectations of environmental interest groups nor the EIA critics' fears of unreasonable costs were proven to be correct. In recent years, EIA experts have viewed the policy with a certain 'Ernüchterung' ('soberness').[45] Two problems, however, remain the subject of legal discussions. Firstly, there has been no 'Popularklagerecht' (i.e. a legal right of appeal for the general public) to challenge decisions in all project categories which has meant that statutory consultees were restricted to those who tended to pursue economic (self-) interests. In order to establish a right of appeal, some environmental interest groups have resolved to purchase land affected by the project. However, not all environmentalists have had the financial means to undertake these legal challenges. As a consequence, many controversial projects were completed without proper EIA. Another problem has been the burden of proof for opponents of projects: planning permission could only be annulled if there was a 'konkrete Möglichkeit' ('definite possibility') that a formal EIA would have produced condemning evidence necessitating planning refusal. In 1997 the Federal Administrative Court in Berlin (Bundes-Verwaltungsgericht or BVerwG) put another damper on EIA 'sympathetic' court rulings with the result that opponents of 'harmful' projects continue to face major difficulties in EIA enforcement.[46]

In sum, formal *determinants* in the German *layer* such as its federal constitution, its fragmented political-administrative structure and sectoralised legal system hindered the *filtering* process of the 1985 Directive (see figure 6.3) and, indeed, hinder the *filtering* of the 1997 follow-up Directive. Focusing on the 1985 Directive, the policy was in many respects incompatible with German legal tradition and its media-oriented processing of environmental matters. Not surprisingly, the commitment to adjust long-established standards was rather limited. The 'direct effect' of the Directive and the prospects of enforcement was limited also. Partly for this reason, experts began to view the effectiveness of the EIA policy with a certain 'soberness'. Although, German practitioners enjoyed expertise and large resources with environmental policies that set quantitative and qualitative standards (e.g. emission limits), the cross-media EIA policy had a bad start indeed in Germany.

There are clear indications that the above mentioned *determinants* also hinder the *filtering* of the 1997 Directives. German legislators and practitioners have, to a certain extent, adjusted their planning framework and have learned some valuable lessons in EIA. However, there has also been a certain 'soberness' with regards to the 1985 Directive's effectiveness which does not seem to disappear with the 1997 follow-up Directive. Since the 1997 Directive has not been transposed properly, it is difficult to assess at this stage to what extent the new Directive will influence (and improve) German planning practices in future.

Figure 6.3: The German Layer and the EIA Directive

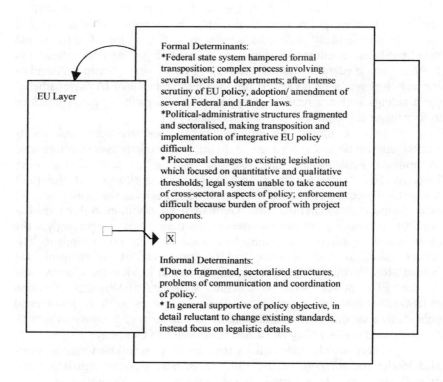

EU Layer

Formal Determinants:
*Federal state system hampered formal transposition; complex process involving several levels and departments; after intense scrutiny of EU policy, adoption/ amendment of several Federal and Länder laws.
*Political-administrative structures fragmented and sectoralised, making transposition and implementation of integrative EU policy difficult.
* Piecemeal changes to existing legislation which focused on quantitative and qualitative thresholds; legal system unable to take account of cross-sectoral aspects of policy; enforcement difficult because burden of proof with project opponents.

X

Informal Determinants:
*Due to fragmented, sectoralised structures, problems of communication and coordination of policy.
* In general supportive of policy objective, in detail reluctant to change existing standards, instead focus on legalistic details.

The Subnational Layer: Bavaria and the EIA Directive

Formal Transposition

Bavaria's handling of the 1985 Directive reflected in many ways Germany's general difficulties with the policy. At the same time, as figure 6.4 illustrates, Bavaria featured 'unique' informal obstacles which made the *filtering* process of the Directive even more difficult. Sharing legislative powers with the Federal level, Bavaria was required to formally transpose parts of the EIA Directive within its boundaries of competency. In particular, Bavaria and other Länder were asked to clarify which authorities were in charge of EIA procedures and provide threshold criteria for Annex II projects which belonged to the Länder level. Until recently, Bavaria followed its obligations only to a certain extent: the Bavarian 'Verordnung' (ordinance) of 20. July 1990 regulated which authority was 'federführend' (in charge) of EIA procedures.[47] In addition, a 1993 'Verordnung' confirmed already established EIA standards for the reparcelling of agricultural land.[48]

In December 1999 Bavaria finally introduced its own EIA law, the 'Bayerische UVP-Richtlinie Umsetzungsgesetz' (BayUVPRLUG), which tackles both 1985 and 1997 EIA Directives but applies only for planning applications processed after 14. March 1999.[49] While Bavaria has been the first Land to take such step, green NGOs have complained already about the (what they perceive to be) 'inadequate' transposition of the policy. StMLU officials, however, have argued that they transposed a 'tolerable amount' of EIA Directive provisions and that they have thereby avoided unnecessary bureaucracy and 'red tape' for businesses in Bavaria.[50] In the light of Bavarian moves towards deregulation, both EIA Directives were certainly not welcome and discussions continue on the 'right balance' between Bavarian deregulation and EU (re-) regulation.[51]

Prior to the adoption of the BayUVPRLUG in 1999, Bavarian legislators failed to meet one important obligation: they did not specify Annex II criteria for projects which were not covered by Federal Laws but fell within Bavarian competencies. Bavarian legislators claimed that measures equivalent to the EIA policy had already been in existence since 1978 and that the formal establishment of EIA criteria was unnecessary.[52] Bavaria's standpoint did not necessarily imply that 'environmentally harmful' projects were not assessed at all, a point made repeatedly by StMLU officials. However, whether or not Bavarian measures were really compliant with the procedural steps of the Directive, remained to be a question for legal clarification by courts at the Bavarian, Federal or European level.

Bavarians have traditionally opposed 'instructions from outside' and, in the case of the EIA policy, have considered the EU input a major infringement on Bavarian affairs. Indeed, one Bavarian state ministry official confirmed that the 1985 Directive clashed with the EU principle of subsidiarity and that planning matters should remain within *sub*national boundaries.[53] Arguably, Bavaria has only recently accomplished its transposition of the EIA policy and still has to follow up some remaining EIA obligations. There have been implementation problems for both formal and informal *determinant* reasons. Formal *determinants* such as a fragmented political-administrative structure and a sectoralised legal system certainly have contributed towards EIA transposition difficulties in Bavaria. However, formal *determinants* alone cannot explain the cumbersome *filtering* process; Bavaria has also been reluctant to follow 'inconvenient and unwelcome EU instructions'. These 'instructions' hindered the EIA policy's implementation right at the initial stage of legal transposition. Bavaria has caught up with EU obligations in the meantime, however, as environmental NGOs would point out, only to a certain extent.

Practical Implementation

Since the BayUVPRLUG applies to project applications submitted *after* 14. March 1999, it would be too premature to assess its overall impact on planning practices in Bavaria. Focusing therefore on the practical implementation of the 1985 Directive pre-1999, findings do not differ significantly from Germany as a whole. Bavarian EIA experts, too, focused on the legal and technical implications of the EU Directive on planning procedures in Bavaria.[54] While German and Bavarian findings are similar, there are distinctly Bavarian characteristics which have shaped the EIA policy in Bavaria in their own way. They concern in particular Bavaria's informal resistance to co-ordinate 'unwelcome policies from outside', as well as a lack of communication on the progress and effectiveness of the policy. According to a StMLU official, this lack of co-ordination was partly due to shortage of staff and insufficient resources.[55] However, there were two other, more substantial, reasons for the lack of co-ordination. Firstly, the Federal and Bavarian constitutions' emphasis on the principle of vertical and horizontal checks and balances has resulted in a fragmented, sometimes confrontational, political-administrative system (see formal *determinants* in figure 6.4). Despite the Bavarian StMLU's comparatively strong and central role in the environmental policy area, officials from other sectoral ministries and local governments have used the principle to resist StMLU intervention on EIA.

Secondly, in the light of the 1990s' economic pressures (i.e. the recession, increased competition from Central and Eastern European neighbours, pressures associated with the convergence criteria of EMU) Bavarian officials at all levels feared the costs and economic constraints of the EIA Directive and were therefore reluctant to co-ordinate the implementation of the policy. Instead, Bavarians reacted to the economic conditions more rigorously than the Federal level (and other Länder). Their investment and deregulation measures clashed with the objective of the EIA Directive and, more recently, with the 1997 Directive which, after all, seeks to bind planners at all government levels to a tighter framework of 'minimum-regret-planning'.[56] The new BayUVPRLUG was therefore formulated in a way that would accommodate a convenient compromise between meeting EU environmental obligations and allowing for economic (self-) interests.

Returning to the first stage of the 1985 Directive pre-1999, Bavarian planning authorities considered the list of EIA project categories provided by the 'Deutsche Gesetzgeber' ('German legislator') as sufficient. In other words, legislators at the Federal level 'identified EIA projects for them' by providing planners with 'water-tight' threshold criteria. Most projects requiring EIA under the UVPG concerned roads, waste management (disposal and processing) and pipelines (for gas, water and oil). However, by focusing on the UVPG text only, Bavarian planning authorities effectively excluded many Annex II projects which were not specifically mentioned in the Federal and Bavarian laws. These included projects which still required clarification at the Bavarian level and projects whose transboundary impacts were difficult to measure. Furthermore, Bavarian planners also focused on thresholds and thereby avoided case-by-case checks of other possible (accumulative) environmental impacts of projects. To add to this, Bavarian State government officials were preoccupied with the formulation of policies which allowed planning authorities to derogate from EIA project categories and concentrate on 'exceptional cases' only.[57] In combination, the exclusion of certain potentially harmful projects and the proposed derogation measures could only result in a decreasing number of EIAs conducted in Bavaria.

Bavarian planning authorities had no complaints concerning the quality of environmental statements. Developers were acquainted with EIA obligations and often commissioned professional environmental consultants with the production of statements. In fact, in order to avoid obstacles in the planning process, developers often provided too many project details which in turn contributed to the heavy work-load of planners. In contrast, environmental interest groups complained about the inadequate and biased content of environmental statements and demanded the inclusion of questions on the wider (even global) environmental implications of projects. The developers' study of two or three project alternatives was considered insufficient. Instead, environmentalists wished to discuss 'moral' questions whether to tolerate and promote economic development at all in a sustainable society. Planning authorities preferred less time-consuming 'simplified' approaches, focusing either on one proposal and its alternative or on a 'zero sum calculation' (i.e. the study of the environment before and after a project was completed). In many cases, environmentalists put enough public pressure on developers and planners to adopt a wider perspective.[58] Bavarian planners could use the 1997 Directive as a means to set the 'right' scoping framework for each planning application requiring EIA. But it is more likely that planners, developers and environmentalists will continue with their elaborate discussions on what items exactly should be considered for each EIA.

    In terms of information and consultation, the 1985 Directive hardly made a difference in Bavaria. Guidelines on information and consultation procedures derived from existing Federal Laws on environmental protection and pollution control as well as administrative regulations.[59] In general, project applications were made public through official notices, the local press and *Amtsblätter* (German equivalent for *gazette*). 'Affected' parties and the public had four weeks time to comment on project applications. Despite the fact that existing Federal legislation allowed only indirect participation for environmental interest groups, Bavarian planning authorities were conscious of the 'green' pressure and therefore involved as many parties as possible.[60] In some cases, Bavarian environmental interest groups which enjoyed large memberships and financial resources, bought property affected by planning applications and thereby created a legal right to be consulted. Some of them even succeeded with their legal challenges as administrative courts confirmed their claims. Consequently, Bavarian planning authorities accepted environmental interest groups as a 'necessary evil' and an influential force in the planning process and adopted a more approachable attitude towards consultees and the public in general.

    According to Bavarian planning officials, Federal and Bavarian laws (in particular the Federal Emission Law of 1974) had more of an environmental influence on planning decisions than the EIA Directive.[61] This attitude has not really changed with the BayUVPRLUG of 1999. Many planning officers stated that the EIA policy influenced decisions only in so far as additional costs, 'unnecessary' work and delays occurred. For instance, the planning permission for the nuclear research station 'München II' was delayed by nine months as a result of EIA requirements.[62] On the other hand, Bavarian planning officers could not deny that the 'inconvenient' extension of environmental investigation generated relevant evidence for consideration. Overall, however, Bavarian planning officers stressed that the policy did not significantly change planning behaviour and planning decisions.

Enforcement

In terms of enforcement, Bavaria's administrative courts appeared more 'sympathetic' towards the 1985 Directive. In particular, they attempted to shift the burden of proof and required planning authorities to successfully *deny* that formal EIA compliance would lead to a different planning decision. They also considered the EIA Directive a step towards widening the legal right of appeal. However, in 1997 the Federal Administrative Court in Berlin rejected this 'sympathetic' interpretation.[63] Consequently, project opponents in Bavaria have faced (and continue to do so) serious difficulties indeed in challenging planning decisions on EIA grounds.

In sum, Bavarian implementors took a lukewarm and pragmatic view of the 1985 Directive. Incompatibilities between the Directive and Bavaria's formal *determinants* (i.e. its administrative structures, its legal system and environmental standards, see figure 6.4) rendered the implementation of the EIA policy difficult. With the exception of Bavarian environmentalists, EIA practitioners in Bavaria also demonstrated informal resistance over the policy's implementation. While many planning authority officials agreed that an EU-wide harmonisation of planning standards (and EIA standards in particular) was important, the same officials also believed that Bavarian standards were already set at a high level and that the policy was not really necessary in Bavaria. In fact, the EIA policy caused 'inconvenient' and 'avoidable' work for practitioners. When asked about the 1985 Directive, one planning official complained about the 'flood of legislation coming from the EU'.[64] Similar comments were made more recently regarding the 1997 Directive and, indeed, other EU environmental policies. These complaints about too much 'eurocracy' confirm the Bavarians' general attitude to pursue their own policies without disturbances from 'outside'. All Bavarian officials who replied to research enquiries stated that they had not been consulted by EU actors or the Federal Government on the 1985 Directive and its 1997 amendment. They had little interest in consultation with (and guidance from) Federal and EU actors. Instead, Bavarian planners were confident in their own EIA expertise and were interested only in information exchange with other partner authorities.

The 1985 Directive (and, indeed, the 1997 Directive) faced major obstacles during the *filtering* process in Bavaria. Even the German term for EIA, 'Umweltverträglichkeitprüfung' (UVP), was criticised by Bavarian officials. UVP implied that projects could only be accepted if negative effects on the environment were fully absorbed. According to StMLU officials, this interpretation raised false hopes which could not be fulfilled.[65] For Bavarian officials, the Federal UVPG of 1990 made one difference: EIA formalities were expensive and time-consuming.[66] These burdens were not welcome at a time when Bavaria attempted to cut 'red tape' in planning after the 'fat years' of economic success.[67] Campaigns such as the 'initiative for the speeding up of planning procedures in Schwaben' contributed towards the trend for more lenient environmental standards.[68] This trend, however, ran counter the Commission's efforts to harmonise environmental planning standards further. While the 1997 Directive faced major implementation problems, Bavarians have signalled already their opposition against the strategic environmental assessment Directive, which is due to be adopted by the Council of Ministers and the EP by the end of 2001.

Figure 6.4: The Bavarian Layer and the EIA Directive

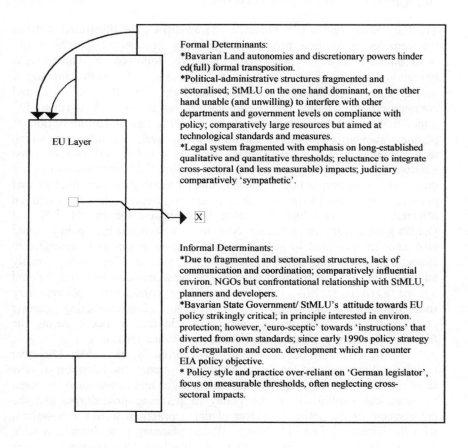

EU Layer

Formal Determinants:
*Bavarian Land autonomies and discretionary powers hinder ed(full) formal transposition.
*Political-administrative structures fragmented and sectoralised; StMLU on the one hand dominant, on the other hand unable (and unwilling) to interfere with other departments and government levels on compliance with policy; comparatively large resources but aimed at technological standards and measures.
*Legal system fragmented with emphasis on long-established qualitative and quantitative thresholds; reluctance to integrate cross-sectoral (and less measurable) impacts; judiciary comparatively 'sympathetic'.

X

Informal Determinants:
*Due to fragmented and sectoralised structures, lack of communication and coordination; comparatively influential environ. NGOs but confrontational relationship with StMLU, planners and developers.
*Bavarian State Government/ StMLU's attitude towards EU policy strikingly critical; in principle interested in environ. protection; however, 'euro-sceptic' towards 'instructions' that diverted from own standards; since early 1990s policy strategy of de-regulation and econ. development which ran counter EIA policy objective.
* Policy style and practice over-reliant on 'German legislator', focus on measurable thresholds, often neglecting cross-sectoral impacts.

Conclusion: Subnational Regions Play a Key Role in the Implementation of the EIA Directive

The Case Study on the EIA Directive (85/337/EEC) has illustrated in detail the complex and often problematic *filtering* process of a typical EU environmental policy. The EIA Directive followed a cumbersome bargaining process which involved a wide range of actors pursuing a variety of (conflicting) interests through formal and informal communication channels (see *feed-back* arrows in *map*). Yet, (then) EC policy-makers entered new territory by agreeing that a harmonisation of planning standards at national and *sub*national levels was necessary which took account of environmental impacts. Policy-makers followed environmental as well as economic level-playing-field motivations in their quest for a common policy. However, this harmonisation also implied that planning competencies in the national and *sub*national *layers* would be affected, an aspect which was considered unwelcome by the UK and Danish governments in particular. Moreover, a common EIA policy would also have the potential to restrict economic development and consequently clash with economic self-interests on the ground. The policy finally adopted 1985 was therefore a compromise between considerations for and against a common EIA policy and provided considerable discretionary room for national and *sub*national implementors. The conflicting interests at the outset of the bargaining process would later re-occur during the *filtering* of the policy through the national and *sub*national *layers*.

In order to be *filtered* through properly, the 1985 Directive required adjustments of legal-administrative systems, the adoption of new laws which would set the policy into the national and *sub*national contexts, the practical application of the policy in planning procedures, and the enforcement of the policy whenever planning practices were not compliant with the Directive. The EIA policy affected planning procedures in a wide range of project categories and therefore involved all government *layers* and their actors throughout the implementation process. The case study has illustrated how actors in all *layers* perceived, interpreted and accommodated the policy according to their particular formal and informal circumstances. In the *filtering* process, national and *sub*national actors either tended to keep the policy's impact to a bare minimum, were eager to fill legal gaps (and misinterpreted the policy, see Germany's comprehensive project list), welcomed the policy as a tool to support sustainable development (see Scottish planners), or they perceived the policy as an 'invasion' to their own competencies (see StMLU and

Bavarian planners). Considering the divergent conditions and attitudes, it comes as no surprise that the 1985 Directive was *filtered* differently in each *layer*.

Although the national *layers* set the initial tone for implementation, it was mainly within the *sub*national regions that planning officials *filtered* the policy further by implementing and applying the EIA policy on the ground. In the process, the *sub*national regions featured a number of implementation similarities with their 'mother' states but also some distinctly Scottish and Bavarian *determinants* which influenced significantly the EIA policy in practice. It is therefore important to distinguish between national and *sub*national *layers* and study them separately in order to gain a more accurate and comprehensive picture of the whole EU environmental policy process.

The implementation of the 1985 Directive, first of all, depended on formal *determinants* inside the *layers*. With regard to the *sub*national regions, their constitutional position in the Member States as well as their legal traditions and their internal political-administrative structures influenced the policy. In the case of Scotland, the transposition of the 1985 Directive was comparatively swift and uncomplicated; even though Scotland required a Statutory Instrument separate from the rest of the UK. Scottish Office officials simply followed their DoE colleagues and adopted the Directive verbatim. The Scottish Office SI left many discretionary powers of planning authorities untouched but also confirmed the Secretary of State's ultimate power in determining final planning decisions. The 1997 Directive has demonstrated that despite devolved powers and more careful scrutiny at the Scottish level, transposition and implementation practices have not changed significantly since the setting up of the Scottish Parliament and Executive. It has to be said, though, that the 1997 Directive did not require substantial changes in the planning framework and did not necessitate intensive scrutiny by Scottish legislators. It is therefore possible that with a more radical and less acceptable EU environmental Directive the transposition and implementation process in Scotland may not be as smooth in future.

In contrast to the UK and Scotland, the Federal and Bavarian transposition and implementation of the 1985 Directive (and, indeed, the 1997 Directive) was far from smooth. The political-legal systems compelled legislators to transpose the 1985 Directive through a complicated and, to a certain extent, controversial process. While UK and Scottish implementors transposed a bare minimum of the EIA policy without considering further clarifying provisions (they thereby avoided legal disagreements with the Commission), their Federal and Bavarian counterparts scrutinised and interpreted the policy in detail, creating a fragmented and technocratic policy framework. In certain areas (such as Annex II projects) this framework departed from the original EIA policy and resulted in disputes with the Commission over the Federal (and Bavarian) laws' compliance with the Directive. To date these disputes have not been resolved and are likely to continue in the near future with the 1997 Directive. Having highlighted the fragmented legal-administrative system as a major obstacle, the federal system and particularly Bavaria's position as a Land with large autonomous powers have also meant that the Free State could adopt a separate BayUVPRLUG in 1999 which in many ways sets a precedence over EIA legislation at the Federal level and, indeed, in other Länder. In this sense, the federal system has facilitated EU environmental policy implementation at least in one of the *sub*national *layers*.

Overall, the case study has demonstrated how formal *determinants* such as the constitutional position and the legal systems of the *sub*national regions can shape their ability to implement EU environmental policies. Examining only formal *determinants* would therefore suggest that as long as constitutional and legal settings are favourable, EU environmental policy implementation should be smooth and successful. But the implementation of the 1985 Directive was also shaped by informal *determinants* such as the *sub*national regions' policy styles, attitudes, priorities and relationships. In Scotland, the EIA Directive was not welcomed with open arms. In fact, Scottish Office officials made no secret of their scepticism towards the policy, partly because the policy was seen as a 'red tape' brake on economic development in Scotland. It was therefore left to the discretion of individual planning officers to decide whether or not to apply the EIA policy effectively. Some planners considered the policy as a useful tool from EU policy-makers to integrate environmental considerations more forcefully, while many others ignored the policy as much as possible for other (economic) priorities.

The Bavarians, too, were reluctant to restrict economic development in their communities, especially at a time when the recession (and other pressures) hit them hard. However, in Bavaria's case the key informal obstacle in the implementation of the EIA Directive could be found in the Bavarians' attitude towards 'instructions from outside'. In particular, the EU was seen as an 'illegitimate' policy-making level in the planning policy area. Bavarian officials were (over-) confident in their own standards and their extensive environmental expertise and resources. They traditionally focused on quantitative and qualitative standards and resisted integrative environmental policies such as the EIA Directive. As a result, Bavarian State officials have only recently (and reluctantly) responded to EIA legal obligations. Considering all factors, a complex (and inter-dependent) mix of formal and informal *determinants* in the *sub*national *layers* have shaped EIA policy implementation performances in Scotland and Bavaria (confirming Arguments 1 and 2).

Scotland and Bavaria have featured divergent political-administrative systems (in the Scottish case an 'evolving' system) with different formal and informal *determinants*. Yet, despite these differences, their EIA policy implementation outcomes were strikingly similar. In both regions, the EIA policy had a moderate impact on planning practices and the key objective (i.e. a level-playing-field in minimum-regret-planning) was not fully achieved. The case study has carved out two reasons for this shared disappointing result. Firstly, EU environmental policies are enormously complex and often involve policy objectives which are based on environmental and economic considerations. In order to be realised, EU environmental policies require the full commitment of key implementors. In the case of the EIA Directive of 1985, commitment was required from legislators at the national and *sub*national levels, planning officers and, to a certain extent, developers and the public (in particular environmental NGOs and other consultees). This commitment, however, was often missing from planning officials and developers. The study has demonstrated that EU environmental objectives such as 'environmentally sensible' minimum-regret-planning can later clash with implementors' more immediate economic and administrative considerations on the ground. The economic argument for the harmonisation of planning standards at the outset of the process may have been plausible for actors on the ground, but when it comes to economic (self-) interests, *sub*national implementors in particular have demonstrated a protectionist attitude towards their local economies. The 1997 Directive has sought to establish a stronger 'level-playing-field' for all. However, whether or not the new Directive is clear and binding enough to impose EU-wide discipline on actors who have economic (self-) interests in mind remains to be seen.

The study has also revealed a certain 'euro-scepticism' towards 'instructions from outside' among *sub*national implementors, an attitude which can only dampen the pursuance of EU environmental policies. *Sub*national regions have been important in the process, yet they have been almost absent from EU policy-making which produces legislation they then are left to implement. Obviously, the exchange of views on the EIA policy practice (indicated in the *map* with *feed-back arrows*) has been rather limited between the *layers*. It remains to be seen whether the more recent EU environmental policy strategies of 'partnership' and 'dialogue' can consolidate this (perceived) gap between EU policy-makers and *sub*national implementors and thereby contribute to more acceptable and effective EU environmental policies. The new strategies have not prevented more controversies surrounding the adoption and implementation of the 1997 Directive. Reactions from national and particularly *sub*national actors during and after the policy's adoption signalled that the new EIA policy was perceived as another 'EU-imposed' constraint on planning policies and economic objectives.[69] The concept of partnership and dialogue therefore appears to be ineffective in closing the gap between EU environmental policy-makers' intentions and the implementors' 'reality'. These difficulties in establishing a well-functioning dialogue between EU policy-makers and implementors contribute towards (and are part of) a wider debate currently occupying politicians, the media and the public on the future of the EU as an ever widening, efficient but also democratic and legitimate Union.

The case study has provided detailed evidence on the EU environmental policy implementation deficit by focusing on *sub*national actors and their policy performances on the ground. The final Chapter draws together the key findings of the research. It re-addresses the key arguments, assesses the usefulness of the *multi-layered implementation map* for further investigations and, finally, presents an outlook of the future of the EU environmental policy.

Notes

[1] See *O.J.* (1985).

[2] Among others see Wood (1995) for a detailed account and discussion of the EIA Directive.

[3] See in particular Hien (1997); Jessel (1997); Macrory (1994); Wood and Jones (1997).

[4] Wood (1995) provides a list of earlier, more ambitious, EIA policy proposals.

[5] In full 'Council Directive 97/11/EC amending Council Directive of 27 June 1985 on the assessment of the effects of certain public and private projects on the environment (85/337/EEC)'.

[6] Statutory Instruments are legally binding but do not require approval by parliament. The EIA Directive's implementation deadline was 3. July 1988; the UK formally implemented the Directive on 15. July 1988.

[7] For England and Wales see Statutory Instruments *The Town and Country Planning. Assessment of Environmental Effects Regulations 1988* (SI 1988 No.1199); *The Town*

and Country Planning (Assessment of Environmental Effects) (Amendment) Regulations 1990 (SI 1990 No.367); *The Town and Country Planning (Assessment of Environmental Effects) (Amendment) Regulations 1994* (SI 1994 No.677).

[8] For Scotland see *Scottish Circular* 15/99 and *Planning Advice Note* (PAN) 58; For England and Wales Regulations see SI No. 293; see also DETR *Circular* 2/99.

[9] See in particular Commission (1993); Wood and Jones (1991); Treweek at al (1993); Lee and Colley (1992).

[10] For a more recent study see Wood and Jones (1997).

[11] See Scottish Office (1990); DoE (1989); DoE (1994a); and Wood and Jones (1991).

[12] See Alder (1993); Ward (1993).

[13] For instance, the planning permission procedure was extended from 8 weeks to 16 weeks to allow for the new information and consultation obligations.

[14] While the SI outlined the items of Annex III of the EIA Directive in detail (in Schedule 3), the document more likely to be consulted was the Circular. It provided inadequate Annex III information: '[t]his statement must include a description of the project; a description of the measures envisaged in order to avoid, reduce and if possible remedy significant adverse effects; the data required to identify and assess the main effects which the project is likely to have on the environment; and a non-technical summary of this information.'

[15] Other bodies included the Health and Safety Executive and the Scottish Environmental Protection Agency (SEPA).

[16] See Scottish Office (1990).

[17] For Scottish Executive views see *Planning Bulletin* (1999b); for StMLU views see *Pressemitteilung* (1999).

[18] Smith used a research format developed and applied by the EIA Centre, University of Manchester. For his study, Smith selected 25 environmental statements submitted to the Scottish (Office) Development Department between 1988 and 1990. In addition, Smith conducted 14 interviews with planning authorities, developers, consultants and consultees in Scotland.

[19] For an appraisal of high standards in North Sea oil environmental assessment and management in Scotland, see Nelson and Butler (1993).

[20] Quotations from a telephone interview with the Scottish Office official, 5. April 1995, Edinburgh.

[21] One Scottish Office official stressed that the advice given was only 'indicative', i.e. it was for the local authorities to apply the criteria mentioned in the Scottish Office Circular. Written correspondence, 4. December 1997, Edinburgh.

[22] One planning official stressed that the style and content of EIAs depended upon his 'own requirements'. Written correspondence, 21. April 1997, Perth.

[23] Lack of experience was mentioned by one local authority (Linlithgow, 17. April 1997), lack of resources by another local authority (Aberdeen, 28. May 1997).

[24] Planning officer, interview, 16. June 1997, Glasgow.

[25] See Smith (1990); also interest groups commented on inadequate environmental information being given during planning procedures at the 'People and the Environment: A Common Cause' Conference, held by Scottish Wildlife and Countryside Link, 14. February 1997, Perth.

[26] Only one planning officer stated that two or three environmental statements were returned. Telephone interview, 2. June 1997, Lochgilphead.

[27] In the case of the motorway M74 Northern Extension, the following bodies were

consulted: Historic Scotland, Strathclyde Passenger Transport Executive, Royal Fine Arts Commission, Coal Authority, Scottish Power, British Gas, Clydeport, Railtrack, The Mineral Valuer, British Telecom, Clyde Calders Project, Strathclyde Police, Scottish Office Environment Department, Historic Scotland and the Scottish Wildlife Trust. Written correspondence, 23. July 1997, Glasgow.

[28] EIA information contributed towards moderate changes of the M74 Northern Extension application (Planning Officer, interview, Glasgow, 16. June 1997); EIA information contributed towards the refusal of two wind farm applications (Planning Officer, telephone interview, 2. June 1997, Lochgilphead).

[29] The EIA policy 'did not make a big difference' when two applications for sewage treatment plants were refused planning permission, according to one planning officer. Written correspondence, 31. March 1997, Elgin.

[30] In an EIA 'related' case (the case was in progress at the time of the 1985 Directive's formal transposition, the policy was therefore not legally binding), the M77 'Road Route Extension' was pushed through the planning process by the Scottish Office, Strathclyde Regional Council and Kilmarnock District Council. The latter planning authority was particularly interested in housing development and trade links generated by the new motorway. Strathclyde Regional Council official, interview, 7. August 1995, Glasgow; Strathclyde Regional Councillor, interview, 29. August 1995, Glasgow.

[31] See Scottish Executive's PAN 58 (1999).

[32] For further information see Williams (1991).

[33] The same planning officers, however, also pointed out the policy's weaknesses which required improvement. For instance, one planning officer stated that EIAs and environmental statements 'struggled to gain public confidence' because they were not seen to be 'impartial, being funded and on occasions prepared by applicants'. Written correspondence, May 1997, Melrose.

[34] One planning officer stated that the 1985 Directive was not really necessary and that he was not interested in further guidance from the Scottish Office and the EU unless these contacts led to a reduction of work. Written correspondence, May 1997, Elgin.

[35] A Scottish Executive official confirms that it is too early to judge the practical implications of the 1997 Directive and that an assessment has not yet been conducted. Written correspondence, 4. June 2001.

[36] English translation: assessing the natural environment's 'ability to absorb' projects and their damaging impacts.

[37] The sixteen Federal Laws included areas such as emission control, nuclear safety, infrastructure, and nature conservation. For a detailed list see Vedder (1990).

[38] See *Gemeinsames Ministerialblatt* (1995).

[39] See 'Rede von Bundesumweltminister Jürgen Trittin anlässlich der Sitzung des Dt. Bundestages am 5. April 2001', BMU website.

[40] The separate paragraphs in the UVPG referred to changes of other Federal Laws at a later stage and provided transitional derogation measures which would apply until the laws in question were amended.

[41] Annex II categories not mentioned in the UVPG list included the manufacture and assembly of motor vehicles and manufacture of motor vehicle engines; storage facilities for petroleum, petrochemical and chemical products; industrial estate development projects. For a detailed list see Commission (1993).

[42] The Commission initiated another infringement proceeding against the German Government for failing to fully comply with the EIA Directive. While German

legislators are busy transposing the new Directive, the 'old issue' of Annex II categories is currently being investigated by the ECJ. If the Court agrees with the Commission, Germany could pay a fine of €237,600 per day until the project categories are included in German law. See UVP-Netz website.

[43] See Weber and Hellmann (1990).

[44] For German studies on EIA practices see Commission (1993); Schwab (1997); Kollmer (1994 and 1995); see also UVP-Netz website.

[45] For an assessment of the EIA enforcement practice in Germany see Hien (1997).

[46] For further details on the BverwG's reaction to Länder court rulings, see Bavarian enforcement section below.

[47] See *Bayerisches Gesetz- und Verordnungsblatt* (1990).

[48] See *Allgemeines Ministerialblatt* (1993).

[49] See *Gemeinsames Verwaltungsblatt* (1999).

[50] See StMLU *Pressemitteilung* (1999a).

[51] See Hösch (2001).

[52] See StMLU (1995).

[53] See Wegner (1993).

[54] For Bavarian studies see in particular Vedder (1990); Weber (1993); Seidel (no date).

[55] Bavarian Ministry official, written correspondence, 7. January 1997, Munich.

[56] For a summary of Bavaria's deregulation policy see Böhm-Amtmann (1998).

[57] For instance, one Bavarian proposal concerned the speeding-up of road building projects. The Bavarian state government argued that the 'old' Länder should adopt the same derogation measures as the 'new' Länder. These temporary measures were intended to lift the 'new' Länder economies to the 'old' Länder level. See Viebrock (1992).

[58] For instance, the project 'Franken II' (power station extension) was delayed following demands for more information by the public. Planning permission was eventually given after intense public scrutiny. See Seidel (no date).

[59] See in particular Paragraph 73 of the VerwverfG, also Paragraph 29 of the Federal Nature Protection Law.

[60] For instance, the Schwaben planning authority involved 46 parties in the consultation process of Bundesautobahn A8; among them local communities, farmers associations, environmental agencies, private sector representations, national heritage societies and environmental interest groups. See Regierung von Schwaben (1996).

[61] Among others, officials from the StMLU (19. August 1996), Schwaben (May 1997), Niederbayern (5. June 1997) and Oberbayern (24. April 1997) stated that the EIA hardly had an influence on planning decisions.

[62] StMLU official, written correspondence, 24. July 1997, Munich.

[63] One well-publicised example was the Bavarian administrative court (VGH) ruling of the B15neu motorway (15. February 1996). The court dismissed the Bavarian state government's decision to permit the B15neu on the grounds that a proper EIA had not been conducted. In April 1997 the BVerwG dismissed the ruling but also referred the case back to the VGH for further consideration of other legal aspects. A final decision on the B15neu was still in progress at the time of writing. Bund Naturschutz in Bayern e.V., written correspondence, 30. July 1996, Landshut. See also Bund Naturschutz e.V. website.

[64] Planning official, written correspondence, 4. June 1997, Schweinfurt.

[65] Another point of criticism was the measuring of cross-media impacts, culminating in the question how cultural goods and fauna (two areas mentioned in the EIA Directive) could

possibly have an impact on each other. StMLU officials, interview, 19. August 1996, Munich.
[66] Regional authorities such as Oberfranken complained about additional work and costs associated with the EIA policy. Written correspondence, 26. March 1997, Bayreuth.
[67] As one StMLU official commented: 'die fetten Jahre sind vorbei' (the good days are over). Interview, 19. August 1997.
[68] See Regierung von Schwaben (1996a).
[69] For further details see, UVP-Gesellschaft website.

References

Alder, J. 'Environmental Impact Assessment. The Inadequacies of English Law', *Journal of Environmental Law*, vol.5, No.2, pp.203-220.

Allgemeines Ministerialblatt, (1993), 'Bayerische Vollzugsrichtlinie Umweltverträglichkeitsprüfung in der Flurbereinigung', 6. August, p.1044.

Bayerisches Gesetz- und Verordnungsblatt (1990), 'Verordnung zur Bestimmung der federführenden Behörde und ihrer Aufgaben gemäß Paragraph 14 Absatz 1 des Gesetzes über die Umweltverträglichkeitsprüfung', Nr.14.

Böhm-Amtmann, E. (1998), *Cooperative Enforcement: Successes, Impediments, Solutions*, Conference Paper, United States Environmental Protection Agency, Washington, 21. January.

Bundesministerium für Umwelt, Naturschutz und Reaktorsicherheit (BMU) (2001), *Rede von Bundesumweltminister Jürgen Trittin anlässlich der Sitzung des Dt. Bundestages am 5. April 2001*, website.

Commission (1993), *Commission Report on the Implementation of Directive 85/337/EEC on the assessment of the effects of public and private projects on the environment*, Com (93) 28 final - vol.13.

DETR (1999), *Town and Country Planning (Environmental Impact Assessment) (England and Wales) Regulations 1999*, SI No. 293.

DETR (1999), *Circular 2/99*.

DoE (1988), *The Town and Country Planning. Assessment of Environmental Effects Regulations 1988*, SI 1988 No.1199.

DoE (1989), *Environmental Assessment. A Guide to the Procedures*.

DoE (1990), *The Town and Country Planning (Assessment of Environmental Effects) (Amendment) Regulations 1990*, SI 1990 No.367.

DoE (1994), *The Town and Country Planning (Assessment of Environmental Effects) (Amendment) Regulations 1994*, SI 1994 No.677.

DoE (1994a), *Environmental Assessment. Evaluation of Environmental Information for Planning Projects. A Good Practice Guide*.

Gemeinsames Ministerialblatt (1995), 'Allgemeine Vorschrift zur Ausführung des Gesetzes über die Umweltverträglichkeitsprüfung (UVPGVwV) vom 18. September 1995', Nr.32, pp.671-694.

Gemeinsames Verwaltungsblatt (1999), Nr.28/99.

Hien, E. (1997), 'Die Umweltverträglichkeitsprüfung in der gerichtlichen Praxis, *Neue Verwaltungsrecht Zeitung*, Heft 5, pp.422-428.

Hösch, U. (2001), 'Das bayerische Gesetz zur Umsetzung der UVP-Richtlinie', *Neue*

Zeitschrift für Verwaltungsrecht, Nr.5.

Jessel, B. (1997), *Die Umweltverträglichkeitsprüfung auf dem Prüfstand*, Bayerische Akademie für Naturschutz und Landschaftspflege, Presse Information Nr.17, 25 April.

Kollmer, N. (1994), 'Die verfahrensrechtliche Stellung der Beteiligten nach dem UVP-Gesetz', *Neue Verwaltungsrecht Zeitung*, Heft 11, pp.1057-1061.

Kollmer, N. (1995), 'Der öffentliche Anhörungstermin im UVP Verfahren (Paragraph 9 UVPG)', *Bayerische Verwaltungsblätter*, 1. August, pp.449-453.

Lee, N. and Colley, R. (1992), *Reviewing the Quality of Environmental Statements*, Occasional Paper 24, EIA Centre, University of Manchester.

Macrory, R. (1994), 'Environmental Assessment and the 'direct effect' doctrine', *ENDS Report*, No.228, January, pp.44/45).

Nelson, J.G. and Butler, R.W. (1993), 'Assessing, planning and management of North Sea oil development effects in the Shetland Islands', *Environmental Impact Assessment Review*, vol.13, No.4, July.

Regierung von Schwaben (1996), *Neubau der Ortsumfahrung Gundelfingen - Lauingen der Bundesstraße 16*, Planfeststellungsbeschluss vom 28. November.

Regierung von Schwaben (1996a), *Schwabeninitiative Beschleunigung von Genehmigungsverfahren*.

Schwab, J. (1997), 'Die Umweltverträglichkeitsprüfung in der behördlichen Praxis', *Neue Verwaltungsrecht Zeitung*, pp.428-435.

Scottish Executive (1999), *Scottish Circular* 15/99.

Scottish Executive (1999a), *Planning Advice Note* (PAN) 58.

Scottish Executive (1999b), *The Planning Bulletin: December 1999*.

Scottish Office (1990), *Environmental Assessment - a guide*, 6/90.

Seidel, R. (no date). *UVP bei Industriestandorten am Beispiel eines Kraftwerk-Genehmigungsverfahrens nach BImschG*.

Smith, J. A. (1990), *Critical Appraisal of the Performance of the Environmental Assessment (Scotland) Regulations since their Introduction*, MSc Dissertation, University of Stirling.

StMLU (1995), *Information Umwelt und Entwicklung in Bayern: Die Umweltverträglichkeitsprüfung*, 1/95.

StMLU (1999), *Staatsminister Dr Werner Schnappauf: Bayern fordert von der EU Umweltmindeststandards und weniger Verfahrensvorschriften*, Pressemitteilung, 21. May.

StMLU (1999a), *Staatsminister Dr Werner Schnappauf zu dpa lby 059 vom 14. April 1999 (Umweltverträglichkeitsprüfung)*, Pressemitteilung.

Treweek J.R. et al (1993), 'Ecological Assessment of Proposed Road Developments', *Journal of Environmental Planning and Management*, vol.36, No.3, pp.295-307.

Vedder, E. (1990). 'Der aktuelle Stand der UVP-Gesetzgebung in der Bundesrepublik Deutschland und in Bayern', *Inhalte und Umsetzung der Umweltverträglichkeitsprüfung (UVP)*, Laufener Seminarbeitraege 6/90, Akademie für Naturschutz und Landschaftspflege, pp.32-35.

Viebrock, J. (1992), 'Beschränkungen der UVP in der Verkehrswegeplanungs-beschleunigung', *Neue Zeitschrift für Verwaltungsrecht*, 11. Jahrgang, pp.939-942.

Ward, A. (1993), 'The Right to an effective Remedy in European Community Law and Environmental Protection: A Case Study of UK Judicial Decisions concerning the Environmental Assessment Directive', *Journal of Environmental Law*, vol.5, No.2, pp.221-244.

Weber, A. and Hellmann, U. (1990), 'Das Gesetz über die Umweltverträglichkeitsprüfung (UVP-Gesetz)', *Neue Juristische Wochenschrift*, Heft 27, pp.1625-1633.

Weber, J. (1993), *Environmental Planning as a Part of Urban Planning in the Federal Republic of Germany - The City of Munich as an Example*, Paper, Universidade Nova de Lisboa.

Wegner, H. A. (1993), 'Die Umweltpolitik der EG im Spannungsfeld zwischen Harmonisierungszwang und Subsidiaritätsprinzip', *Berichte der Bayerischen Akademie für Naturschutz und Landschaftspflege* Sonderdruck, Nr.17.

Williams, R. (1991), 'Direct Effect of EC Directive on Local Authorities', *The Cambridge Law Journal*, vol.50, Part 3, pp.382-384.

Wood, C. (1995), *Environmental Impact Assessment: A Comparative Review*, Longman, Essex.

Wood, C. and Jones, C. (1991), *Monitoring Environmental Assessment and Planning*, DoE Planning Research Programme.

Wood, C. and Jones, C. (1997), 'The Effect of Environmental Assessment on UK Local Planning Authority Decisions', *Urban Studies*, vol.34, No.8, pp.1237-1257.

7   Conclusion: Subnational Regions Matter in the Implementation of EU Environmental Policies

Introduction

The purpose of this research was to move away from the analysis of national governments and their role in EU environmental policy-making and highlight instead *sub*national regions and their influence on the success, or failure, of EU environmental policies. In particular, the research has argued that the study of *sub*national regions and their actors can help explain why the EU is suffering from an implementation deficit in the environmental policy area. By distinguishing between national and *sub*national government levels the aim was to contribute new and vital insights to the study of EU environmental policy implementation, insights which have been hitherto neglected by 'state-centrist' analyses.

To help investigate the EU environmental policy process, the research combined and synthesised relevant study areas and approaches into a heuristic framework. The *multi-layered implementation map* introduced in Chapter 2 built on the evidence of existing policy process and implementation studies, and incorporated the complex EU and environmental policy dimensions. The *map* highlighted, and distinguished between, three government levels or *layers* involved in the *filtering* process of EU environmental policies and categorised influential factors into formal and informal *determinants*. The latter distinction helped identify potential obstacles in the implementation path and explained why many EU environmental policies either failed on the ground or took a different shape in the latter stages of policy implementation.

Figure 7.1: The Multi-Layered Implementation Map Revisited

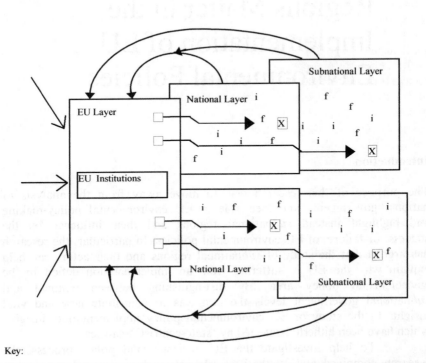

Key:

EU Policy Statement: ☐   EU Policy Target: X

Filtering Process: ⟶   Feed-Back: ↰

External Factors: →

Informal Determinants: i   Formal Determinants: f
* relationships between actors;   * constitutional settings;
* attitudes towards environmental   * political-administrative
  protection and the EU;   structures and resources;
* policy-makers' priorities and   * legal systems and instruments.
  strategies;
* policy styles and practices.

The research confirmed a number of well-publicised studies which highlight an implementation deficit in the EU environmental policy area,[1] a deficit which is arguably more alarming than in other EU policy areas. Indeed, the author's own research evidence (gained from interviews, questionnaires and primary sources) confirmed that at every stage of the implementation process - from the formal transposition and progress report to the enforcement and monitoring of policies - EU Member States have demonstrated difficulties (and often reluctance) in meeting EU environmental obligations. While many studies have contributed valuable insights to the research matter and suggested measures which would solve some of the implementation problems,[2] this study has sought to construct a more comprehensive framework which takes account of specific implementation factors while maintaining an overview of the wider (macro-) context of EU environmental politics. This was primarily done by categorising factors that influence EU environmental policy implementation into formal and informal *determinants* and by distinguishing between EU, national and *sub*national government levels.

The final Chapter addresses the key arguments and then evaluates the *map* in the light of the research findings. It concludes with a short outlook on the future of EU environmental policies and assesses to what extent the EU can act as an environmental 'problem solver'.

Addressing the Key Arguments

Argument (1): Formal *determinants* such as political-administrative structures as well as informal *determinants* such as policy priorities and relationships between actors influence EU environmental policy implementation on the ground. These formal and informal *determinants* are inter-related and cannot be studied on their own.

National and *sub*national actors have processed EU environmental policies in accordance with their particular formal and informal conditions or circumstances. EU environmental Directives have, by their very nature,[3] provided considerable discretion for implementors. They have been formulated in a way that would allow for national and *sub*national variances. While this discretion has been necessary, it has also provided 'loopholes' (discussed in detail in Chapter 3) for implementors whereby policy objectives were either avoided or policies were shaped to fit into implementors' political-administrative frameworks. As a result, many environmental policy *targets* have often been missed.

The study categorised the most influential factors in the implementation process into formal and informal *determinants* shown in figure 7.1. The formal *determinants* referred to:

- constitutional settings
- political-administrative structures and resources
- legal systems and instruments.

The informal *determinants* comprised:

- relationships between actors
- attitudes towards environmental protection and the EU
- policy-makers' priorities and strategies
- policy styles and practices.

Proper implementation of EU environmental policies was often hindered by incompatible legal systems, complicated political-administrative structures or a lack of financial and administrative resources. In addition, informal obstacles such as divergent policy priorities, contradictory policy styles and strategies have often clashed with EU environmental policies. In different combinations or 'mixes', these formal and informal *determinants* have generally rendered the implementation of EU environmental policies difficult.

Chapters 3, 4 and 5 have all demonstrated the extent to which both formal and informal *determinants* have shaped EU environmental policies at different government levels. The inclusion of both formal and informal *determinants* was crucial: an examination of only formal or informal *determinants* alone would have led to inaccurate conclusions. In fact, the Chapters have illustrated how formal and informal *determinants* are inter-related and cannot be studied on their own. It was therefore important to 're-assemble' the formal and informal *determinants* and assess their *combined* influence on EU environmental policies.

If formal *determinants* were to be studied only, the previous centralised state system of the UK and Scotland (i.e. formal constitutional setting) would appear to have been ideal for the implementation of EU environmental policies. Indeed, in comparison with the German and Bavaria, 'instructions from outside' were processed automatically and in a more integrated manner without much political scrutiny (see Chapters 4, 5 and 6). On the other hand, the UK state system featured gaps in co-ordination and co-operation which were particularly evident in the Scottish case. The gaps were mainly due to Scotland's paradoxical position in the centralised state system and associated questions concerning Scottish representation and devolution (see Chapter 5). Therefore the formal constitutional setting of the UK and Scotland had an impact on informal attitudes and relationships between actors which in turn influenced the way in which policies have been processed.

With devolution, again a mix of formal and informal *determinants* influences the policy process. This time the formal setting includes a Scottish Parliament and Executive, which eliminate the somewhat 'hybrid overlap' represented by the Scottish Office and the Secretary of State for Scotland. While this change in formal *determinants* has taken place, informal *determinants* have seen only minor alterations. True, relations between actors from various government levels have been less confrontational, but old communication gaps still exist and may even get worse in the near future. Post-devolution Scotland is still evolving. It would therefore be too early to assess the medium to long-term implications of devolution on informal relationships, priorities and practices. The potential for (radical) informal changes exist thanks to the formal reforms of recent years. However, indications are that the Scots will continue with EU environmental policy business as usual. Given the potential for change, however, they may well some day implement EU environmental policies significantly different from the rest of the UK. For now it is suffice to say that pre- and post-devolution Scotland has featured formal and informal *determinants* that have shaped the *filtering* of EU environmental policies. They have been inter-linked and can therefore not be studied on their own.

In comparison, the complexity of the federal state system has always been problematic for the *filtering* of EU environmental policies in both the German and Bavarian *layers* (see Chapters 4, 5 and 6). Political-administrative structures have been fragmented and provided horizontal and vertical gaps that rendered the co-ordination of policies difficult. In addition, the constitutional checks and balances contributed towards one important informal *determinant*: the general perception that policies 'from outside' should be scrutinised, and if necessary adjusted, at every government level and in every sectoral department. Bavaria in particular fostered this perception with the result that many EU environmental policies, welcome or not, were hindered during the implementation process. However, focusing on formal *determinants* and their informal implications only would have neglected those informal *determinants* in Germany and Bavaria that have eased the implementation of EU environmental policies.

Favourable informal *determinants* included strong public (grass-root) support of 'green' issues and the Federal Governments' campaign for stringent qualitative and quantitative environmental standards (e.g. water quality standards and emission thresholds). Again, formal and informal *determinants* have to be examined together in order to gain an accurate picture of the policy 'reality'.

Exploring informal *determinants* further, until the early 1990s German Governments pursued the idea that stringent environmental standards would provide German producers with a competitive advantage in 'eco-friendly' products and markets. This strategy facilitated the implementation of many EU environmental policies (such as policies on large combustion plants, catalytic converters and lead-free petrol). However, the strategy gave way to an economic policy priority of the late 1990s, which resembled the UK's laissez-faire approach. The Government also moved away from its EU position as a 'green man' demanding the highest environmental standards, and moved instead towards a more lenient environmental policy with 'more affordable' EU *targets*.

This policy has not changed significantly with the new 'red-green' coalition Government under Schröder. Despite the Greens' efforts in introducing more stringent eco-taxes and other 'green' policies, the Government is unlikely to recreate the 'green man' image in the light of continuing economic pressures. The German political system relies too much upon consensus, especially between political parties and interest groups (in other words, formal *determinants* set informal priorities). Moreover, given that the new Federal Chancellor Schröder has to reassure German businesses and voters that economic conditions will improve in the near future, it is very unlikely that the Government will introduce 'deep green' policies which will complement (and improve) EU environmental policy implementation and secure sustainable development in Germany.[4]

In comparison, the formal *determinants* (in particular the centralised state system and the FPTP electoral system) allowed Conservative Governments to pursue a laissez-faire economic policy in the 1980s and 1990s. This policy priority was in line with EU 'voluntary action' policies such as the 'eco-audit' but was less compatible with EU policies specifying and controlling qualitative and quantitative standards (such as water and air emission standards). Although the new Labour Government has pursued some radical environmental policy objectives (particularly in the area of infrastructure and transport, see Chapter 4), substantial changes in attitudes and priorities are not evident and are not expected in the near future. It is therefore unlikely that the Labour

Government will focus on the effective implementation of EU environmental objectives and establish a policy of sustainable development in the UK.

Scotland and Bavaria have featured informal *determinants* that have differed in many respects from their 'mother' states. Arguably, policy priorities and strategies in Scotland and Bavaria have responded more vigorously to changes in public attitudes and priorities on the ground (see in particular Chapters 5 and 6). As a Federal Land, Bavaria has been able to pursue its own policy strategies and conduct policy changes that have been more radical than in other Länder. In this sense, the formal constitutional setting has helped Bavaria to establish its own set of priorities. Having formulated an environmental policy at an early stage with stringent qualitative and quantitative requirements, Bavarians conducted a political U-turn in the early 1990s towards a policy of voluntary action and de-regulation (see in particular the *Umweltpakt* in Chapter 5). This new policy was intended to facilitate and encourage economic development and counter-act trade competition 'threats' from the rest of the EU and Eastern European neighbours. In the light of this strategic change, EU environmental policies, which implied financial and administrative costs as well as restrictions to economic development in Bavaria, have been processed with considerable reluctance if not downright opposition by Bavarian implementors (for Bavarian reactions to EU environmental policy obligations see in particular Chapter 6).

Although Westminster influenced considerably the political process North of the border (e.g. Conservative governments introduced privatisation and de-regulation policies), Scotland featured some distinct informal *determinants*. Priorities in Scotland were by and large conservative: next to economic stability, prosperity and growth, environmental considerations remained low-priority issues for the public and political-administrative actors in Scotland. True, environmental issues were considered increasingly important in the late 1990s. Some important economic sectors in Scotland such as tourism and the wool and whisky industries conveniently combined economic and environmental interests, and a number of environmental initiatives in Scotland demonstrated that the Scots could not ignore environmental matters completely. However, these adjustments were not substantial enough to ensure the proper implementation of EU environmental policies. Scotland's 'paradoxical' structures (discussed in detail in Chapter 5) contributed towards this emphasis on economic priorities. They allowed Scottish local authorities to pursue economic self-interests, while (former Conservative) central governments were able to implement economic policies of de-regulation in Scotland. EU environmental policies were implemented accordingly in Scotland: the Scottish Office tended to transpose a bare minimum of 'costly and inconvenient' EU environmental policies, while local administrators enjoyed considerable discretion when applying the policies on the ground. They either followed the Scottish Office's example and limited - often severely - the impact of EU environmental policies as in the case of the EIA Directive, or they welcomed the 'green' input from the EU as useful environmental tools.

Devolution introduced a new formal *determinant* set-up and although informal relations between actors (i.e. between 'Scottish' and local authority levels) and political debates (i.e. on Scotland's 'destiny' in the UK) have evolved, attitudes towards environmental and economic issues have remained essentially the same. Unless informal *determinants* change substantially (e.g. as a result of a major pollution incident or economic crisis), the *filtering* of EU environmental policies in Scotland will continue to follow an all too familiar pattern. Overall, both Scottish and Bavarian cases have demonstrated how informal interests and formal circumstances are intertwined, influential and distinct from that of their 'mother' states.

It is striking how policy-makers' economic priorities have influenced the implementation process of EU environmental policies in all *layers*. Economic level-playing-field motives which initially compelled EU policy-makers to adopt many EU environmental policies (see Chapter 3 and the EIA Case Study featured in Chapter 6) have tended to evaporate during the latter stages of the EU environmental policy process. Depending on the political parties in power and their strategies towards temporary economic pressures, Member States and *sub*national regions have pursued their own economic advantage, much to the detriment of EU environmental policies which often required economic and financial concessions. In all *layers* informal *determinants* have influenced the way in which EU environmental policies have been implemented. Yet informal priorities and relationships have not developed in a vacuum, they have been shaped by formal constitutional circumstances which have, in turn, influenced the way in which EU environmental policies have been received and processed by implementors on the ground. Again, it is true to say that formal and informal *determinants* are inter-related and cannot be studied on their own.

*Argument (2): Sub*national regions and their actors play a central role in the EU environmental policy process. They shape the implementation of most

EU environmental policies.

This Chapter has highlighted already the importance of investigating *determinants* in the Member States at large as well as *determinants* within the *sub*national regions. Indeed, the separate investigation of EU, national and *sub*national *layers* has provided a more accurate overview of the divergent *determinants* which shape the development and implementation of EU environmental policies. This research compared national and *sub*national conditions and examined in detail the *sub*national regions and their influence on the EU environmental policy implementation process. It concluded that *sub*national regions play a vital role in the success or failure of most EU environmental policies.

EU environmental policies such as Directives setting water quality standards, identifying and protecting areas of environmental interest, and procedural Directives which affect decision-making in policy areas such as planning require implementation at *all* government levels. The policies' success therefore depends upon national *and* *sub*national actors and their capacities and willingness to realise EU environmental objectives. Although the national governments take the initial steps of formal transposition, it is mainly the *sub*national regions which are concerned with any further details of legal transposition and, more importantly, the subsequent practicalities of most EU environmental policies. In practice, *sub*national actors have shaped EU environmental policies decisively; in the implementation process they have accommodated (or failed to accommodate) the policies into their own political-administrative frameworks. This research has emphasised that *sub*national actors do not live in a vacuum but are influenced by circumstances (i.e. formal and informal *determinants* described above) which differ in many respects from the circumstances which shape environmental politics in the Member States at large. The detailed study of *sub*national regions and their implementation performances is therefore of vital importance for an accurate picture of the EU environmental policy practice.

Chapters 5 and 6 provided detailed evidence of Scotland and Bavaria's 'unique' implementation conditions and performances with EU environmental policies. In the Scottish case, it was admittedly a central government department (i.e. a national *layer* institution), the Scottish Office, which transposed the bulk of EU environmental policies into the *sub*national context. However, the Scottish Office processed the policies separately from the rest of the UK and took into account Scottish characteristics and interests such as its infrastructure and certain economic sectors such as the wool and fishing industries. In this sense, the Scottish

Office already influenced EU environmental policies in a way which was 'distinctly Scottish'. For instance, the EIA Directive of 1985 was transposed comparatively swiftly by Scottish Office officials and followed the DoE example, yet the tone of the Scottish Office policy documents departed in many respects from the documents applying to the rest of the UK. In comparison, Scottish Office officials were almost apologetic about the 'inconveniences' caused by the EU environmental policy. This reaction reflects a general reluctance to restrict economic development for the sake of 'European' environmental objectives in a region which was (and still is) trying to come to terms with its peripheral disadvantage.

While the Scottish Office played a significant role in the way EU environmental policies were processed in Scotland, the effectiveness of EU environmental policies such as the EIA Directive depended upon individuals' attitudes and priorities on the ground. In other words, Scottish local authority officials had large discretionary room over the policies' application (unless, of course, the Secretary of State intervened) and implemented EU environmental policies according to their own local priorities and practices. In the EIA case, some Scottish practitioners welcomed and applied the Directive as a long-overdue instrument which enhanced environmental planning in Scotland, while many others followed their economic priorities and feared the costs, economic restrictions and administrative burdens associated with the Directive. Overall, the implementation of EIA Directive depended upon the *sub*national *layer* (i.e. Scotland) and its actors who *filtered* the policies in line with formal and informal *determinants*. Although the Scottish Office was in a somewhat hybrid position between national and *sub*national *layers*, the *sub*national *layer's* role in the overall *filtering* process of EU environmental policies was nevertheless striking. This finding has not been altered with the new Scottish Parliament and Executive. In fact, devolution has eliminated the overlap between *layers* and will, as the Scottish political system will evolve further, contribute to a clearer and 'distinctly Scottish' implementation picture.

Bavaria's handling of EU environmental policies has been different in many respects. Firstly, as the EIA case has illustrated, Bavaria has had considerable formal difficulties in transposing EU environmental policies due to the fragmented federal state system and sectoralised legal framework. However, in contrast with the Federal level (and indeed with other Federal Länder) Bavarian political-administrative actors have shown considerable reluctance in complying with 'inconvenient and unnecessary' EU standards. Bavarians have resisted adjustments to their legal-administrative standards and procedures as much as possible. Environmental policies that affected other policy sectors (such as the EIA Directive) as well as policies with environmental *targets* and thresholds different from Bavarian standards have faced informal obstacles in Bavaria. In the EIA case, StLMU officials followed their obligations by initially transposing only parts of the Directive (in the form of two ordinances) and thereby demonstrated their opposition against 'inconvenient instructions from outside'. In principle, Bavarians have supported objectives similar to EU environmental policies. Bavarians also have had the financial and administrative means (and the know-how) as well as a strong environmental lobby to support EU environmental policies. However, recent policy practice has shown that EU environmental policies have been confronted with both a strategic change towards environmental policy 'lenience' *and* a 'typically Bavarian' resilience against instructions from outside. In comparison with Germany at large, the Bavarians' 'euro-scepticism' has been striking. Political-administrative actors have been confident in their own standards and challenged the legitimacy of the EU as a policy-making level in this policy area. These Bavarian characteristics have made the *filtering* process incredibly difficult.

Comparing the two *sub*national regions, the research has identified and highlighted substantial formal and informal *determinant* differences between Scotland and Bavaria. Yet despite these differences, the regions' implementation performances have been strikingly similar. One immediate conclusion is that the success of EU environmental policies has depended upon the 'mix' of *determinants* and that EU policy-makers have to accept that even the 'best tailored' or 'smartest' (to use Ingram and Schneider's term) policies face complex *determinant* combinations and are therefore most likely to 'get stuck' during the *filtering* process. Both Scottish and Bavarian *layers* have featured complex mixes which included favourable *determinants* facilitating policies as well as unfavourable *determinants* hindering the policies' implementation (and vice versa). For instance, the strong environmental lobby in Bavaria, which has been very much in favour of EU policy objectives, has not been able to compensate for the serious difficulties in pressing especially cross-sector policies into a fragmented and often confrontational political-administrative system. Similarly, in the case of Scotland the most integrative (and centralised) formal structure has not prevented informal reluctance by political-administrative actors whose policy priorities clashed with EU environmental policies.

Another explanation for the similar implementation performances can be found in certain informal *determinants* which both regions share. True, Scotland and Bavaria have been different in terms of constitutional settings, political-administrative structures and policy styles and strategies. But the research has also highlighted that economic considerations as well as tensions between government *layers* have mattered considerably in both Scotland and Bavaria. As far as the economic imperative is concerned, the regions' strategies and policy styles have differed over the years, but the paradoxical relationship between economic and environmental interests has proven to be a dominant factor in the pursuance of EU environmental policies (confirming the 'environmental dimension' argument in Chapters 1 and 2). By and large, environmental objectives have been (and continue to be) perceived as contradicting economic interests on the ground. Often, environmental and economic considerations which have motivated Member State governments to adopt EU environmental policies have not been *filtered* through to (or shared by) actors at the *sub*national level. As a result, EU environmental policies have lacked the necessary support by implementors on the ground.

In the case of Scotland and Bavaria, EU environmental policies have been particularly susceptible to economic considerations: both regions have shown a more protectionist attitude towards their local economies than the national governments for whom a European level-playing-field appeared to be a more immediate, plausible and desirable policy objective. Scotland has had to compensate for its peripheral disadvantage in the Single Market, while Bavaria has had to deal with increased competition from its Eastern European neighbours who have attracted businesses away from Bavaria. The UK and Germany at large have faced these economic pressures as well, but not to such a decisive extent as their *sub*national regions. Comments such as −

> the fat years are over, we cannot afford stringent environmental standards any longer (Bavarian StLMU official)

and −

> we don't want any more environmental policies from Europe, the most endangered species in the Highlands is man (Scottish business representative)

illustrate the *sub*national regions concern over EU environmental policy restrictions on economic developments.[5]

As far as tensions between government levels are concerned, both regions have shown gaps in communication and co-ordination between government levels. For different formal and informal reasons (i.e. the impact of Scotland's previously 'paradoxical' but now devolved position in the UK state system and Bavaria's 'verflechtet' yet confrontational position in a federal state system), actors in both regions have had internal communication difficulties as well as weak external links with national and EU actors. This was reflected mainly in terms of weak formal contacts and informal attitudes towards each other which often culminated in scepticism, mistrust and conflict. Relations between the two *sub*national regions and the other *layers* have not been favourable. Indeed, both Bavarians and Scots openly complained about 'inconvenient' and 'unwanted' policies from the EU. This suggests a general dissatisfaction of *sub*national actors over their restricted access and involvement in the 'making' of EU policies which affect their competencies considerably.

One obvious solution to the perceived communication gap and poor record of implementation would be to adjust the EU policy-making process so that *sub*national regions and their important role in the overall process is taken into account. In other words, the rapport between EU, national and *sub*national actors (indicated with *feed-back arrows* in figure 7.1) should be strengthened. To a certain extent, the EU and the Commission in particular, have already recognised the problem and have attempted to consolidate the *layers* by introducing 'partnerships' and 'dialogue groups' (such as IMPEL). Indeed, recent Treaties, *Environmental Action Programmes*, secondary legislation and other EU environmental initiatives (outlined in Chapter 3) have attempted to involve *sub*national actors more closely in EU environmental policy-making and thereby commit them more strongly to an environmental level-playing-field. These attempts, however, have thus far yielded only limited success. They have not prevented new criticism over EU environmental policy proposals (such as the EIA follow-up Directive of 1997 and the SEA Directive) and have not solved old problems of communication and co-ordination between actors when dealing with EU environmental policies. In other words, formal adjustments of the Treaties, *Environmental Action Programmes* and environmental laws have not (yet) altered informal relationships between actors, policy priorities and attitudes towards environmental protection.

Apart from the attempts of involving *sub*national actors in EU environmental policy-making, *sub*national regions themselves have to a certain extent contributed towards closer co-operation between the *layers*. They have succeeded with some of their demands for more political competency at the national and EU levels. As a result, Scotland now enjoys a stronger presence in Brussels with Scotland House, while Bavaria has succeeded with its campaign for increased and formalised *sub*national participation at the EU level. With Scottish devolution and Scotland House and more formalised EU decision-making powers for Bavaria, both regions may enjoy new (and greater) influence on EU policy-making which in turn could result in EU environmental policies more acceptable to these regions. However, this trend towards government 'fusion' could also make relationships between political-administrative actors more complicated which could result in 'confusion' and wider gaps of communication in future.[6] More importantly, with increased participatory powers and a stronger 'national' identity and self-confidence, *sub*national regions may now oppose policies or adjust them to their liking with the result that one of the EU's key objectives, the economic and environmental level-playing-field, may be ignored completely. While this 'worst case scenario' has not yet appeared, it remains to be seen to what extent the two divergent developments of regionalisation and European integration (which includes the implementation of, and compliance with, common policies) can be accommodated so that common environmental policies can be implemented effectively and successfully in every part of the EU. Already, Bavarian political-administrative actors have shown more interest in EU policy-making links and markedly less interest in EU policy implementation links. *Sub*stantial improvements in relationships between the *layers* and their actors are therefore unlikely.

The potential conflict between economic and environmental interests and the tensions between government levels constitute problem areas for the EU environmental policy process which cannot be eliminated completely by the EU's current environmental strategies. Arguments such as:

- energy efficiency measures cut financial costs for businesses
- 'green' technologies are lucrative in the long-term
- 'partnerships' and 'dialogue' groups between government levels will contribute towards better performances

have had some impact on attitudes and behaviours on the ground. They have appealed predominantly to the voluntary commitment of political-administrative actors, businesses and citizens. However, they have not brought about the substantial changes necessary for the successful and effective implementation of EU environmental policies. Implementation outcomes will therefore continue to be strikingly similar and often disappointing. Considering the multitude of obstacles and resistance by many implementors on the ground (particularly when economic and environmental interests clash), the implementation of EU environmental policies and sustainable development in general will remain a tough and long-term objective in Europe.

An Assessment of the Map and its Usefulness for Future Investigations

The research has demonstrated that the study of *sub*national regions in the EU environmental policy process is important for a more refined as well as comprehensive analysis. By focusing on the 'third' government level, the research has followed an approach similar to that of 'conventional' multi-level governance studies. However, in contrast to studies by Marks (1996) and Hooghe (1996), this research has not argued that *sub*national actors are (increasingly) on an equal footing with actors from EU and national levels, especially during EU policy-making. Rather, it has used the 'multi-level' perspective to investigate the involvement of three government levels and to highlight *sub*national regions as political arenas where most of EU environmental policy implementation takes place. The research has used a *multi-layered implementation map* which describes EU environmental policy implementation as a *filtering* process whereby EU environmental policies have to go through various *layers* before they reach their actual implementation *target*. The term *layers* (highlighting the government levels involved in the process) has been particularly useful; it has helped to illustrate that the EU and national levels are not the only levels that shape EU environmental policy implementation. In fact, applied to the study of Scotland and Bavaria, the *map* has helped to demonstrate that *sub*national regions and their actors have played a vital role in the process. Moreover, with the help of the *map*, the research has uncovered 'mixes' of *sub*national *determinants* which have been distinct from their 'mother' states and which have shaped significantly the implementation of EU environmental policies.

With the *multi-layered implementation map*, the research has not presented a new theory or model able to predict EU environmental policy outcomes. Rather, the *map* has provided a guidance tool which helps identify implementation *determinants* without losing sight of the overall policy process and the government levels involved. The *map* can be applied for the study and comparison of other *sub*national regions and is not restricted to Scotland and Bavaria. Similarly, the *map* can be used for policy areas other than the environment and could generate similar findings on other implementation deficits in EU policy areas such as transport where the subnational regions' role is equally strong.[7] Yet, the *map* is particularly suitable for the environmental policy area because it tends to involve a vast range of inter-related yet conflictual aspects (i.e. formal and informal *determinants*) which may not be found to the same extent in other EU policy areas (e.g. cohesion policy) where interests between the main actors are (arguably) clearer or more harmonious. EU environmental policies also tend to require major adjustments and concessions, often involve compromises between a multitude of actors who pursue conflicting interests (i.e. adjustments of informal *determinants*), and finally know neither time nor geographical boundaries. In addition, EU environmental policies affect actors at all levels: actors from EU institutions, national government representatives and experts, *sub*national administrators, local communities and interest groups. The *map* encapsulates (and distinguishes between) government *layers* and implementation *determinants* and is therefore particularly useful and suitable for the study of EU environmental policies.

There are, however, shortfalls with the *map*. The *map* (and the research itself) resembles an 'old-fashioned' *top-down* approach whereby EU environmental policy implementation is investigated in a linear manner and focuses mainly on obstacles along the way. While this may be true to a certain extent, the author nevertheless has been careful to take account of the whole policy process and other 'directions' that contribute to the dynamism and complexity of EU environmental policy.[8] In addition, the research has not proposed an 'instructivist' solution to implementation problems as some *top-downers* would suggest. 'Instructivism' would ignore implementors' legitimate interests in their own affairs and conditions as well as the 'real' problems in dealing with EU environmental policy obligations. By the same token, the research has avoided a *bottom-up* approach similar to that of Elmore's 'backward mapping' (1979) where implementors' *feed-back* (see *feed-back arrows* in figure 7.1) plays the most dominant role in the whole process. Indeed, Elmore argues that policy-makers should assess implementors' abilities, attitudes and resources first before they even consider a new policy (see Chapter 2). In contrast to Elmore, this research has argued that implementors' interests and behaviours should not be accepted as unchangeable or a yardstick by which future policies should be measured. Future EU environmental objectives need not be set at a low, pragmatic level just because it is convenient for economic and political-administrative actors at the national and *sub*national levels. *Backward mapping* would ultimately defeat the legitimacy of the EU environmental objectives themselves (they are supposed to tackle pollution and prevent further deterioration) and the legitimacy with which EU policy-makers have adopted the policies. After all, EU policy-makers have been 'instructed' (though indirectly) by their electorate and interest groups to pursue common policies.[9]

There appears to be a third option which seeks to combine *bottom-up* and *top-down* approaches. This option describes EU politics as a complex and dynamic system of European governance where national (and *sub*national) policies are influenced by the EU and vice versa.[10] While this third option is useful in so far as it highlights the 'unique' (and increasing) complexity of EU environmental politics, it does not tackle the immediate problem: the discrepancy between EU environmental policy 'ambitions' and policy 'reality', a 'reality' that has been described by many as disappointing. To this day, EU researchers and practitioners such as Krämer raise the issue of (and complain about) the implementation deficit in EU environmental policy. In order to address the issue, policy practice on the ground has to be investigated and compared with policy objectives as outlined in the Directives and Regulations. This inevitably involves a *top-down* perspective. The research has responded to the much publicised implementation deficit and has sought to shed more light onto the differences in formal and informal conditions (differences that are legitimate and often unavoidable) which render EU environmental policy implementation difficult. Moreover, this research has gone one step further than many other studies by disaggregating national and *sub*national government levels, by investigating implementation *layers* separately and by comparing their differences (and similarities) in formal and informal circumstances. The distinction between government *layers* has helped to illustrate how varied formal and informal circumstances are and how these circumstances have shaped EU environmental policies differently. In this context, the research has shown how *sub*national regions have differed in many respects from their 'mother' states.

Of course, this research is by no means complete. Firstly, the research could have explored in more detail the aspect of *feed-back* from the national and *sub*national *layers* to the EU *layer*. *Feed-back* deserves more attention in the near future considering the recent initiatives of partnership and dialogue which have contributed towards the establishment of groups such as IMPEL. While this research has taken account of some of the national and *sub*national responses to EU environmental policy-making, a specific and direct link between an EU policy and its *feed-back* has not been established. In its defence, it was not the purpose of this research to establish such a link. Nevertheless, implementors' response and its impact on the overall EU environmental policy process is an under-researched area that could benefit from further investigation.

The research could also include more case studies on EU environmental Directives other than the EIA Directive of 1985. On its own, the EIA Directive case study has generated valuable insights that have confirmed the overall concept of *multi-layered* implementation. The case study has illustrated how complex the *filtering* process is, how it involves all *layers* and how 'mixes' of formal and informal *determinants* can influence implementation on the ground. The question remains whether the comparison of different types of EU environmental Directives (or Regulations) would generate similar (or different) findings. This research has defended its choice of case study and has argued that the study of the EIA Directive is sufficient to highlight typical (and unique) problems of EU environmental policy implementation on the ground. Still, a comparison of two or several EU environmental policies could contribute more valuable evidence to this research.

Similarly, the research could be extended to include policy areas other than the environment. This could help assess to what extent EU environmental policy is more complex and difficult to implement than are, for instance, the competition policy or the CAP. There are already studies by Butt-Philip and others that compare the EU environmental policy with other EU policy areas. More comparative studies could confirm and underline the argument that the EU environmental policy area is a problematic one and that EU environmental objectives require more commitment and determination by implementors than in any other policy area.

Focusing on *sub*national regions, further comparison of regions other than Scotland and Bavaria could also be useful. While Scotland and Bavaria were carefully selected for their striking differences as well as similarities (for a justification see Chapters 1 and 2), other regions could contribute aspects of formal and informal *determinant* differences that have not been found in the Scottish and Bavarian *layers*. Again, this research should serve as an encouragement for further investigation.

Finally, the *map* itself has shown limitations during the research. Apart from its tendency to guide researchers *top-down*, the *map* has proven to be too simplistic in some areas, especially when dealing with EU policy tools, institutions and government levels. These weaknesses have been acknowledged in Chapter 2 as minor and unavoidable. The problem of distinguishing between government *layers*, however, was particularly apparent in Chapters 5 and 6. For instance, the Scottish Office could not be confidently ascribed to *one* of the *layers* because of its somewhat hybrid position representing both national and *sub*national interests. Nevertheless, the author was careful to acknowledge this hybrid position and argued that the Scottish Office represented in many respects the 'overlap' between UK and Scottish *layers*. Moreover, both Chapters 5 and 6 have shown that there are government levels *below* the *sub*national level that have shaped EU environmental policy implementation in their own way. Chapter 2 already suggested that adding further *layers* below the *sub*national level could contribute to an *even more* refined picture of EU environmental policy implementation. While this may be the case, the question remains as to how many *layers* should be added to a comparative investigation. This question may be difficult to answer considering that political-administrative structures and responsibilities among *sub*national regions are too diverse to press into one comparative format.

Chapter 2 has pointed out that EU environmental policy implementation is enormously complex and that it is difficult to conceptualise all aspects that constitute the subject matter into one framework (Schumann (1991) described this exercise as 'trying to embrace the whole elephant'). This research and its *map* have confirmed that every conceptual framework has its limitations. It is hoped that this research serves as a starting point for further investigations and that it encourages others to take up some of the themes that still require clarification.

A Wider Outlook

This research has contributed another facet to the complex picture of EU environmental policy. It has illustrated how environmental Directives have been particularly susceptible to political-administrative and, more importantly, economic interests. Despite reassurances from the EU Commission that economic and environmental considerations do not necessarily exclude each other (in fact, for a sustainable development they should merge), the research has confirmed that the (perceived) conflict between these two interest areas is still very much alive and that it renders EU environmental policy implementation incredibly difficult.

Apart from the conflict between economic and environmental interests, the EU has had to deal with a more general dilemma that concerns the balancing act between efficiency and democratic legitimacy. Following the principle of subsidiarity, the EU (and particularly the Commission) has tried to consolidate efficient and effective EU decision-making on the one hand with the views and (democratic) decisions of national, *sub*national and local actors on the other hand. While the principle of subsidiarity as well as the Commission's efforts in minimising EU policies have come some way in addressing the dilemma, the wider problem of accommodating all interests and government levels into a comprehensive and well-functioning environmental policy framework still remains.

In the light of clashing interests and questions surrounding the legitimacy with which the EU level adopts common environmental policies as well as the limited success with which these policies are implemented, the pressing question arises whether there is any point in pursuing EU environmental objectives. Another question would be whether many of these EU environmental policies are not merely statements of good intentions which lack 'real' commitment by EU policy-makers and national/ *sub*national implementors. In response to these questions it has to be pointed out that there *are* compelling reasons for tackling environmental problems at the EU level (e.g. transboundary pollution, see Chapter 1). Apart from major environmental problems, there are also economic considerations: Member States are interested in harmonising standards between environmental 'leaders and laggards' in order to strengthen the Single Market. The fact that many implementors will later ignore the level-playing field consideration behind many EU environmental Directives is another matter. There has been *some* progress in regulating and controlling pollution and protecting areas and species of environmental interest. It could be even argued that environmental conditions in Europe would be considerably worse without the EU's involvement in the policy area. In this sense, the EU is a legitimate and important policy-making level in environmental policy. Nevertheless, the EU environmental policy suffers from, what Weale (1996) would call, an inherent 'pathology': this research has demonstrated that the EU polity is by no means ideal for solving environmental problems because it is so complex and diverse. There are no ideal solutions to fully tackle the above described problems and dilemmas that affect the EU in general and the EU environmental policy in particular. One can only hope that, by shedding more light onto EU environmental policy implementation, policy-makers and implementors will learn from past mistakes and take further steps (however small they may be) to narrow the gap between environmental policy intention and policy 'reality'.

Notes

[1] Among others see Butt Philip (1994); Demmke (1996); Institute for European Environmental Policy (IEEP) (1993); Krämer (1992 and 1997); and various Commission reports such as *Interim Review* (1994).

[2] See in particular Collins and Earnshaw (1992); Butt Philip (1994); and Krämer (1992 and 1997).

[3] EU Directives outline common objectives but the leave the details (i.e. the ways and means) to national and *sub*national implementors. See Chapter 3.

[4] The 'phasing-out' of nuclear energy in Germany is arguably an attempt to (re-) install a 'green man' strategy. This policy, however, has faced severe resistance from the nuclear sector lobby.

[5] The comments were made by an StMLU official in Munich (19. August 1996) and a member of the Scotland Europa Environment Group in Glasgow (26. February 1997).

[6] Rometsch (1995) uses the terms 'fusion' and 'confusion'.

[7] Obviously, if the *map* is applied to other policy areas, the informal *determinant* 'attitudes towards environmental protection and the EU' should be replaced by an equivalent *determinant* such as 'attitudes towards the Trans-European Network'.

[8] For instance, the research has outlined how actors from different government levels and interest groups have sought to influence EU environmental policy-making and how political views and experiences from existing policies have been *fed back* to the EU level.

[9] Hooghe (1998) discusses the legitimacy question and the problem of policy dysfunctionality.

[10] For a discussion of European governance see Weale (1996).

References

Butt Philip, A. (1994), *Regulating the Single European Market: A Comparison of the Implementation of Social and Environmental Legislation*, Research Paper.

Collins, K. and Earnshaw, D. (1992), 'The Implementation and Enforcement of EC Environmental Legislation', *Environmental Politics*, vol.1, No.4, Winter, pp.213-249.

Commission (1994), *Interim Review of Implementation of the EC Programme of Policy and Action in Relation to the Environment and Sustainable Development*, COM (94) 453 final.

Demmke, C. (1996), *Verfahrensrechtliche und administrative Aspekte der Umsetzung von EG-Umweltpolitik*, European Institute of Public Administration, 30. May.

Elmore, R. F. (1979), 'Backward Mapping. Implementation Research and Policy Decisions', *Political Science Quarterly*, vol.94, pp.601-616.

Hooghe, L. (ed) (1996), *Cohesion Policy and European Integration. Building Multi-level Governance*, Clarendon Press, Oxford.

Hooghe, L. (1998), 'EU Cohesion Policy and Competing Models of European Capitalism',

Journal of Common Market Studies, vol.36, No.4, December, pp.457-477.

IEEP (1993), *The State of Reporting by the European Commission in Fulfilment of Obligations contained in EC Environmental Legislation*, London.

Krämer, L. (1992 and 1997), *Focus on European Environmental Law*, Sweet & Maxwell, London.

Marks, G. (ed) (1996), *Governance in the European Union*, Sage, London.

Rometsch, D. (1995), *The Federal Republic of Germany and the European Union. Patterns of Institutional and Administrative Interaction*, University of Birmingham Discussion Papers in German Studies.

Schumann, W. (1991), 'EG-Forschung und Policy-Analyse. Zur Notwendigkeit, den ganzen Elefanten zu erfassen', *Politische Vierteljahresschrift*, 32.Jahrgang, Heft 2, pp.232-257.

Weale, A. (1996), 'Environmental rules and rule-making in the European Union', *Journal of European Public Policy*, vol.3, No.4, December, pp.594-611.

Bibliography

Alder, J. (1993), 'Environmental Impact Assessment - The Inadequacies of English Law', *Journal of Environmental Law*, vol.5, No.2, pp.203-220.

Allgemeines Ministerialblatt (1993), 'Bayerische Vollzugsrichtlinie Umweltverträglichkeitsprüfung in der Flurbereinigung', 6. August, p.1044.

Ashworth, G. (1992), *The Role of Local Government in Environmental Protection. First Line Defence*, Longman, Essex.

Baker, S., Yearley, S. and Milton, K (eds) (1994), *Protecting the Periphery. Environmental Policy in Peripheral Regions of the European Union*, Frank Cass, Ilford.

Baker, S., Kousis, M., Richardson, D. and Young, S. (eds) (1997), *The Politics of Sustainable Development. Theory, Policy and Practice within the European Union*, Routledge, London.

Bakkenist, G. (1994), *Environmental Information. Law, Policy and Experience*, Cameron May, London.

B.A.U.M. Consult (1997), *Die umweltbewusste Gemeinde*, (Survey) May.

Bayerische Staatskanzlei (1995), *Umweltpakt. Miteinander die Umwelt schützen* , pamphlet.

Bayerischer Städtetag (1991), *Bayerischer Städtetag. Aufgaben, Organisation, Mitglieder*.

Bayerisches Gesetz -und Verordnungsblatt (1990), 'Verordnung zur Bestimmung der federführenden Behörde und ihrer Aufgaben gemäß Paragraph 14 Absatz 1 des Gesetzes über die Umweltverträglichkeitsprüfung', Nr.14.

Bayerisches Landesamt für Umweltschutz (LfU) (no date), *Im Dienste des Umweltschutzes*, pamphlet.

Bayerisches Staatsministerium für Landesentwicklung und Umweltfragen (1994), *Umweltschutz in Bayern '94*, pamphlet.

Bayerisches Staatsministerium für Landesentwicklung und Umweltfragen (1995), *Information Umwelt und Entwicklung in Bayern: Die Umweltverträglichkeitsprüfung*, pamphlet, 1/95.

Bayerisches Staatsministerium für Landesentwicklung und Umweltfragen (1996), *Information: Umweltschutz und Landesentwicklung in Bayern*, pamphlet, 10/96.

Bayerisches Staatsministerium für Landesentwicklung und Umweltfragen (1996), *Die Umweltbewusste Gemeinde. Leitfaden für eine nachhaltige Kommunalentwicklung*, pamphlet.

Bayerisches Staatsministerium für Landesentwicklung und Umweltfragen (1997), *Bayern mahnt Pflichterfüllung beim Umweltrecht an*, Presseerklärung

Nr.126/97, 28. Februar.

Bayerisches Staatsministerium für Landesentwicklung und Umweltfragen (1999), *Plan-UVP verursacht neue Prüfpflichten, mehr Zeit- und Kostenaufwand,* press release.

Bayerisches Staatsministerium für Landesentwicklung und Umweltfragen (1999), *Staatsminister Dr Werner Schnappauf: Bayern fordert von der EU Umweltmindeststandards und weniger Verfahrensvorschriften, Pressemitteilung,* 21. May.

Bayerisches Staatsministerium für Landesentwicklung und Umweltfragen (1999), *Staatsminister Dr Werner Schnappauf zu dpa lby 059 vom 14. April 1999 (Umweltverträglichkeitsprüfung),* Pressemitteilung.

Bayerisches Staatsministerium für Landesentwicklung und Umweltfragen (2001), *Umweltschutz in der Europäischen Union – Anspruch und Wirklichkeit,* press release.

Bayerisches Staatsministerium für Landesentwicklung und Umweltfragen (2001), *Trittin operiert mit falschen Zahlen,* press release.

BBC (1997), *Scotland Today,* TV news report, (16. June).

Berman, P. (1980), 'Thinking about Programmed and Adaptive Implementation: Matching Strategies to Situations', in H. M. Ingram and D. E. Mann (eds), *Why Policies Succeed or Fail* Sage, London, pp.205-227.

Blowers, A. (1987), 'Transition or Transformation? - Environmental Policy under Thatcher', *Public Administration,* vol.65, Autumn, pp.277-294.

Böhm-Amtmann, E. (1998), *Cooperative Enforcement: Successes, Impediments, Solutions,* Conference Paper at the United States Environmental Protection Agency, Washington 21-22. January.

Böhmer-Christiansen, S. A. and Skea, J. (1991), *Acid Politics: Environment and Energy Policies in Britain and Germany,* Belhaven Press, London.

Böhmer-Christiansen, S. A. (1990), 'Emerging International Principles of Environmental Protection and their Impact on Britain', *The Environmentalist,* vol.10, No.2, pp.96-112.

Bomberg, E. (1994), 'Policy Networks on the Periphery: EU Environmental Policy and Scotland', *Regional Politics & Policy,* vol.4, No.1, Spring, pp.45-61.

Bomberg, E. and Peterson, J. (1996), *Decision-making in the European Union. Implications for central-local government relations,* Joseph Rowntree Foundation, York.

Bomberg, E. and Peterson, J. (1999), *Decision-making in the European Union,* MacMillan, Basingstoke.

Bressers, H. et al (1994), 'Networks as Models of Analysis: Water Policy in Comparative Perspective', *Environmental Politics* vol.3, No.4.

Bullmann, U. (1997), 'The Politics of the Third Level', in C. Jeffery (ed), *The Regional Dimension of the European Union: Towards a Third Level in Europe?,* Frank Cass, London, pp.3-19.

Bund Naturschutz in Bayern e.V. (no date), *Der Bund Naturschutz informiert:*

Verkehrspolitik. Totalschaden!, pamphlet.

Bund Naturschutz in Bayern e.V. (1989), *BN-Position. Das Alpenprogramm des Bundes Naturschutz*, pamphlet.

Bund Naturschutz in Bayern e.V. (1993), *BN-Position. Ökologisches Landessanierungsprogramm. Stellungnahme zur Fortschreibung des Landesentwicklungsprogramms Bayern*, pamphlet.

Bund Naturschutz in Bayern e.V. (1993), *Rettet die Donau! Stoppt die Kanalisierung!*, pamphlet.

Bund Naturschutz in Bayern e.V. (1995), *BN-Position. Zukunft für die Landwirtschaft. Aktualisierte agrarpolitische Forderungen des BN*, pamphlet.

Bund Naturschutz in Bayern e.V. (2000), *Bund Naturschutz bekräftigt Unterstützung des Volksbegehrens 'Macht braucht Kontrolle'*, press release PM14/LGS, (May).

Bundesgesetzblatt (1990), 'Gesetz zur Umsetzung der Richtlinie des Rates vom 27. Juni 1985 über die Umweltverträglichkeitsprüfung bei bestimmten öffentlichen und privaten Projekten (85/337/EWG)', Teil I, Nr.6.

Bundesministerium für Umwelt, Naturschutz und Reaktorsicherheit (BMU) (2001), *Rede von Bundesumweltminister Jürgen Trittin anlässlich der Sitzung des Dt. Bundestages am 5. April 2001*, website.

Butt Philip, A. (1994), *Regulating the Single European Market: A Comparison of the Implementation of Social and Environmental Legislation*, Research Paper.

Butt Philip, A. and Baron, C. (1988), 'UK Report' in H. Siedentopf and J. Ziller (eds), *Making European Policies Work. The Implementation of Community Legislation in the Member States*, Sage, London.

CEBIS (no date), *Environmental Legislation and Policy for the Manager*, pamphlets.

Central Regional Council (1994), *Environmental Charter*, pamphlet.

City of Edinburgh District Council (1995), *Environmental Strategy*, Department of Strategic Services, pamphlet.

Collins, K. and Earnshaw, D. (1992), 'The Implementation and Enforcement of EC Environmental Legislation', *Environmental Politics*, vol.1, No.4, Winter, pp.213-249.

Comfort, L. K. (1980), 'Evaluation as an Instrument for Educational Change', in H. M. Ingram and D. E. Mann (eds), *Why Policies Succeed or Fail*, Sage, London, pp.35-57.

Commission (1990), *'1992' The Environmental Dimension. Task Force Report on the Environment and the Internal Market*.

Commission (1993), *Report from the Commission on the Implementation of Directive (85/337/EEC)*, COM (93) 28 final.

Commission (1994), *The Commission Eleventh Annual Report to the European Parliament on monitoring the Application of Community Law – 1993*, COM

(94) 500 final.

Commission (1994), *Interim Review of Implementation of the European Community Programme of Policy and Action in Relation to the Environment and Sustainable Development 'Towards Sustainability'*, November, COM (94) 453 final.

Commission (1996), *Proposal for a European Parliament and Council Decision on the Review of the European Community Programme of Policy and Action in Relation to the Environment and Sustainable Development. Towards Sustainability*, January 1996. COM (96) 647 final.

Commission (1996), *Taking European Environment Policy into the 21st Century. Progress Report on the Fifth Environmental Action Programme.*

Commission (1996), *The Commission Thirteenth Annual Report to the European Parliament on monitoring the Application of Community Law - 1995*, COM (96) 600.

Commission (1996), *How is the European Union protecting our environment?* , pamphlet.

Commission (1997), *Durchführung des Umweltrechts der Gemeinschaft. Mitteilung an den Rat der Europäischen Union und an das Parlament*, Communication to the Council and the European Parliament.

Commission (1999), *The Global Assessment of the Fifth Environmental Action Programme. Europe's Environment. What directions for the future?*, COM (99) 543 final.

Commission (no date), *Europa-Pass für Bayern*, pamphlet.

Demmke, C. (1996), *Verfassungsrechtliche und administrative Aspekte der Umsetzung von EG-Umweltpolitik*, European Institute of Public Administration.

Department for the Environment, Transport and the Regions (DETR) (1999), *Town and Country Planning (Environmental Impact Assessment) (England and Wales) Regulations 1999*, SI No. 293.

Department for the Environment, Transport and the Regions (DETR) (1999), *Circular 2/99.*

Department of the Environment (DoE) (1988), *The Town and Country Planning. Assessment of Environmental Effects Regulations 1988*, SI 1988 No.1199.

Department of the Environment (DoE) (1989), *Environmental Assessment. A Guide to the Procedures.*

Department of the Environment (DoE) (1990), *The Town and Country Planning (Assessment of Environmental Effects) (Amendment) Regulations 1990*, SI 1990 No.367.

Department of the Environment (DoE) (1994), *The Town and Country Planning (Assessment of Environmental Effects) (Amendment) Regulations 1994*, SI 1994 No.677.

Department of the Environment (DoE) (1994), *Environmental Assessment. Evaluation of Environmental Information for Planning Projects. A Good*

Practice Guide.

di Fabio, U. (1998), 'Integratives Umweltrecht. Bestand, Ziele, Möglichkeiten', *Neue Zeitschrift für Verwaltungsrecht*, Nr.4, pp.329-337.

Dinan, D. (1999), *Ever Closer Union. An Introduction to European Integration*, MacMillan Basingstoke.

Edye (1997), *Regions and Regionalism in the European Union*, European Dossier 38, University of North London.

Elmore, R. F. (1979), 'Backward Mapping. Implementation Research and Policy Decisions', *Political Science Quarterly*, vol.94, Winter, pp.601-616.

ENDS Report (1995), 'European Environment Agency gets under way', No.240, January, pp.20-23.

Europe (1997), Nr.6903 (N.S.), 30. January.

Falter, J. (1982), 'Bayerns Uhren gehen wirklich anders. Politische Verhaltens- und Einstellungsunterschiede zwischen Bayern und dem Rest der Bundesrepublik', *Zeitschrift für Parlamentsfragen*, Nr.13, pp.504-521.

Freestone, D. (1991), 'EC Environmental Policy and Law', *Journal of Law and Society*, 18(1), Spring, pp.135-154.

Friends of the Earth Scotland (1997), 'Scottish Parliament - Green for Go?', *What on Earth*, Issue 18, Summer, p.6.

Friends of the Earth Scotland (1998), 'View from the Mill', *What on Earth*, Issue 20, Spring, p.3.

Gemeinde Hunding (1995), *Gemeinde Hunding Arbeitsergebnisse: Kommunale Strukturpolitik*, pamphlet.

Gemeinde Kirchdorf im Walde (no date), *Landschaftsplanung Gemeinde Kirchdorf im Wald*, pamphlet.

Gemeinsames Ministerialblatt (1995), 'Allgemeine Vorschrift zur Ausführung des Gesetzes über die Umweltverträglichkeitsprüfung (UVPGVwV) vom 18 September 1995', Nr.32, pp.671-694.

Gemeinsames Verwaltungsblatt (1999), Nr.28/99.

George, S. (1990), *An Awkward Partner: Britain in the European Community*, Clarendon Press, Oxford.

Goldin, I. and Winters, L.A. (eds) (1995), *The Economics of Sustainable Development*, Cambridge University Press, Cambridge.

Gordon, J. (1994), 'Environmental Policy in Britain and Germany: Some Comparisons', *European Environment*, vol.4, part 3, June, pp.9-12.

Grossman, G. M. (1995), 'Pollution and growth: what do we know?', in I. Goldin and L. A. Winters (eds), *The Economics of Sustainable Development*, Cambridge University Press, Cambridge, pp.19-46.

Haigh, N. (1984), *EEC Environmental Policy and Britain. An Essay and a Handbook*, Environmental Data Services, London.

Harper, Robin (2001), 'Spreading the environment too thinly', *Press and Journal*, (23. March).

Heritier, A. (1993), 'Policy-Netzwerkanalyse als Untersuchungsinstrument im europäischen Kontext', *Politische Vierteljahresschrift*, Sonderheft 'Policy-Analyse. Kritik und Neuorientierung', 24/93, pp.432-449.

Heritier, A. (ed) (1994), *Die Veränderung von Staatlichkeit in Europa. Ein regulativer Wettbewerb. Deutschland, Großbritannien, Frankreich*, Leske + Budrich.

Heritier, A. and Knill (1996), 'Neue Instrumente in der europäischen Umweltpolitik: Strategien für eine effektive Implementation', in G. Lübbe-Wolf (ed), *Der Vollzug des Europäischen Umweltrechts*, Erich Schmidt Verlag, Berlin, pp.209-233.

Hien, E. (1997), 'Die Umweltverträglichkeitsprüfung in der gerichtlichen Praxis', *Neue Verwaltungsrecht Zeitung*, Heft 5, pp.422-428.

Hill, J. S. and Weissert, C. S. (1995), 'Implementation and the Irony of Delegation: The Politics of Low-level Radioactive Waste Disposal', *The Journal of Politics*, vol.57, No.2, May, pp.344-366.

Himsworth, C. and Munro, C. (1998), *Devolution and the Scotland Bill*, W. Green and Son, Edinburgh.

Hinterstoißer, F. (1996), 'Umweltpolitik in Bayern - Folgen für die Landwirtschaft', *Landwirtschaft und Umweltpolitik*, Nr.30, pp.25-33.

Hösch, U. (2001), 'Das bayerische Gesetz zur Umsetzung der UVP-Richtlinie', *Neue Zeitschrift für Verwaltungsrecht*, Nr.5.

Hooghe, L. (ed) (1996), *Cohesion Policy and European Integration. Building Multi-level Governance*, Clarendon Press, Oxford.

Hooghe, L. (1998), 'EU Cohesion Policy and Competing Models of European Capitalism', *Journal of Common Market Studies*, vol.36, No.4, December, pp.457-477.

Inglehart, R. (1990), *Culture Shift in Advanced Industrial Society*, Princeton University Press.

Ingram, H. and Schneider, A. (1990), 'Improving Implementation through framing smarter Statutes', *Journal of Public Policy*, vol.10, pp.67-88.

Institute for European Environmental Policy (1993), *The State of Reporting by the European Commission in Fulfilment of Obligations contained in EC Environmental Legislation*, London, November.

Jänicke, M. (1996), *Umweltpolitik der Industrieländer. Entwicklung - Bilanz – Erfolgsbedingungen*, edition sigma, Berlin.

James, P. (1995), *The Politics of Bavaria - An Exception to the Rule*, Avebury, Aldershot.

Jeffery, C. (1996), 'Towards a 'Third Level' in Europe? The German Länder in the European Union', *Political Studies*, vol.44, No.2, pp.253-266.

Jeffery, C. (1996), 'Sub-National Authorities and European Domestic Policy', *Regional and Federal Studies*, vol.7, No.3, pp.204-219.

Jeffery, C. (1996), *The Emergence of Multi-level Governance in the European Union*, Conference Paper 8eme Colloque International de la Revue 'Politique et management Public.

Jessel, B. (1997), *Die Umweltverträglichkeitsprüfung auf dem Prüfstand*, Bayerische Akademie für Naturschutz und Landschaftspflege, Presse Information, Nr.17, 25. April.

John Wheatley Centre (1995), *Working for Sustainability: An Environmental Agenda for a Scottish Parliament*, Final Report, The Governance of Scotland Project, August.

Jordan, A. (1996), 'The European Union: Rigid or Flexible Decision-Making. From Brussels to Blackpool and Southport. 'Post-decisional Politics' in the European Community', in Hampster-Monk (ed), *Contemporary Studies*, PSA Conference Proceedings, pp.1897-1906.

Jordan, A. (1996), *Implementation Failure or Policy Making? How do we theorise the Implementation of EC Policy at the National and Sub-national Level?*, Working Paper, Centre for Social and Economic Research on the Global Environment, University of East Anglia and University College of London.

Keating, M. and Hooghe, L. (1996), 'By-passing the nation state? Regions and the European Union policy process', in J. Richardson (ed), *European Union Power and Policy-Making*, Routledge, London, pp.216-229.

Kellas, J. (1989), *The Scottish Political System*, Cambridge University Press, Cambridge.

Knill, C. and Lenschow, A. (1998), 'Coping with Europe: the impact of British and German administrations on the implementation of EU environmental policy', *Journal of European Public Policy*, vol.5, No.4, pp.595-614.

Kollmer, N. (1994), 'Die verfahrensrechtliche Stellung der Beteiligten nach dem UVP-Gesetz', *Neue Verwaltungsrecht Zeitung*, Heft 11, pp.1057-1061.

Kollmer, N. (1995), 'Der öffentliche Anhörungstermin im UVP Verfahren (Paragraph9 UVPG)', *Bayerische Verwaltungsblätter*, 1. August, pp.449-453.

Koopmans, R. (1995), *Democracy from below. New Social Movements and the political system in West Germany*, Westview Press, Oxford.

Krämer, L. (1991), 'The Implementation of Community Environmental Directives within Member States: Some Implications of the Direct Effect Doctrine', *Journal of Environmental Law*, vol.3, No.1, pp.39-56.

Krämer, L. (1992), *Focus on European Environmental Law*, Sweet and Maxwell, London.

Krämer, L. (1997), *Focus on European Environmental Law*, Sweet and Maxwell, London.

Landesbund für Vogelschutz (1997), *LBV fordert: Schutzgebietnetz NATURA 2000 erweitern*, Presseinformation, Nr. A31-97, 22. Juli.

Lean, G. (1995), 'Where did all the fresh air go?', *The Independent on Sunday*, The Sunday Review, 5. March, pp.4-9.

Lee, N. and Colley, R. (1992), *Reviewing the Quality of Environmental Statements*, Occasional Paper 24 (2.ed.), EIA Centre, University of Manchester.

Lenaerts, Koen (1994), 'The Principle of Subsidiarity and the Environment in the

EU. Keeping the Balance of Federalism', *Fordham International Law Journal*, vol.17, Part4, pp.846-895.

Liberatore, A. (1991), 'Problems of transnational policymaking: Environmental Policy in the European Community', *European Journal of Political Research*, No.19, pp.281-305.

Lindemann, H. H. and Delfs, S. (1993), 'Vollzug des Europäischen Umweltrechts. Lösungsansätze zur Überprüfung und Verbesserung', *Zeitschrift für Umweltrecht*, Nr.6, pp.256-263.

Lipset, S. M. (1994), 'Binary Comparisons. American Exceptionalism - Japanese Uniqueness', in M. Dogan and A. Kazancigil (eds), *Comparing Nations. Concepts, Strategies, Substance*, Blackwell, Oxford, pp.153-212.

Lothian Regional Council (1994), *Charter for Action on the Environment. 3rd Environmental Action Plan* , pamphlet.

Lübbe-Wolf, G. (1996), 'Stand und Instrumente der Implementierung des Umweltrechts in Deutschland', in G. Lübbe-Wolf (ed), *Der Vollzug des Europäischen Umweltrechts*, Erich Schmidt Verlag, Berlin, pp.77-106.

Macrory, R. (1994), 'Environmental Assessment and the 'direct effect' doctrine', *ENDS Report*, No.228, January, pp.44\45.

Maloney, W. and Jordan, G. (1995), 'Joining Public Interest Groups: Membership Profiles of Amnesty International and Friends of the Earth', in J. Lovenduski and J. Stanyer (eds), *Contemporary Political Studies*, vol.3, PSA, University of York, pp.1137-1153.

Malunat, B. M. (1994), 'Die Umweltpolitik der BRD', *Aus Politik und Zeitgeschichte*, B49/94, 9. Dezember, pp.3-12.

Mann, M. (1995), 'Widespread condemnation of action programme review', *European Voice*, Survey: Environment, 30. May - 5. June, p.17.

Mann, M. (1996), 'EU struggles to find right shade of green', *European Voice,* 30.May-5.June, p.13.

Marks, G. (1993), 'Structural Policy and Multilevel Governance', in A. Cafruny and G. Rosenthal (eds), *The State of the European Community. The Maastricht Debates and Beyond*, vol.2, Lynne Riemer, Harlow, pp.391-410.

Marks, G. (ed) (1996), *Governance in the European Union*, Sage, London.

Marks, G., Hooghe, L. and Blank, K. (1996), 'European Integration in the 1980s: State-Centric v. Multi-Level Governance', *Journal of Common Market Studies*, vol.34, No.3, September, pp.341-378.

Marks, G., Scharpf, F. W., Schmitter, P. C. and Streeck, W. (eds), *Governance in the European Union*, Sage, London.

Mauritz, M. (1995), *Natur und Politik: Die Politisierung des Umweltschutzes in Bayern*, Andreas Dick Verlag, Neustraubling.

McAteer, M. and Mitchell, D. (1996), 'Peripheral Lobbying! The Territorial Dimension of Euro Lobbying by Scottish and Welsh Sub-central Government', *Regional and Federal Studies*, vol.6, No.3, pp.1-27.

McCaig, E. and Henderson, C. (1995), *Sustainable Development: What it means to the general public*, MVA Consultancy for The Scottish Office Central Research Unit.

McCormick, J. (1991), *British Politics and the Environment*, London, Earthscan.

McCormick, J. (1993), 'Environmental Politics', in P. Dunleavy (ed), *Development in British Politics 4*, Macmillan, Basingstoke, pp.267-284.

McDowell, E. (1995), 'The Environmental Movement in Scotland: An Empirical Analysis of Leading Group Activists', in J. Lovenduski, Joni and J. Stanyer (eds), *Contemporary Political Studies*, vol.3, PSA, University of York, pp.1154-1163.

McHugh, F. (1996), 'Voluntary accord get mixed response', *European Voice*, 30. May - 5. June, p.18.

Meny, Y. et al (1996), *Adjusting to Europe. The Impact of the European Union on National Institutions and Policies*, Routledge, London .

Merkel, W. (1999), 'Legitimacy and Democracy: Endogenous Limits of European Integration', in J. Anderson (ed), *Regional Integration and Democracy*, Rownan and Littlefield, pp.45-67.

Midwinter, A., Keating, M. and Mitchell, J. (1991), *Politics and Public Policy in Scotland*, Macmillan, London.

Moravscik, A. (1991), 'Negotiating the Single European Act: National interests and conventional statecraft in the European Community', *International Organisations*, vol.45, pp.19-56.

Moravscik, A. (1993), 'Preferences and Power in the European Community: A Liberal Intergovernmentalist Approach', *Journal of Common Market Studies*, vol.31, No.4, December, pp.473-524.

Müller, E. (1989), 'Sozial-liberale Umweltpolitik. Von der Karriere eines neuen Politikbereiches', *Aus Politik und Zeitgeschichte*, 47-48/89, 17. November, pp.3-15.

Nelson, J. G. and Butler, R. W. (1993), 'Assessing, planning and management of North Sea oil development effects in the Shetland Islands', *Environmental Impact Assessment Review*, vol. 13, No.4, July.

Nugent, N. (1999), *The Government and Politics of the European Union*, MacMillan, Basingstoke.

O'Brien, D. (1980), 'Crosscutting policies, uncertain compliance, and why policies often cannot succeed or fail', in H. M. Ingram and D. E. Mann (eds), *Why Policies Succeed or Fail*, Sage, London, pp.83-106.

O'Toole Jr., L. (1986), 'Policy Recommendations for Multi-Actor Implementation. An Assessment of the Field', *Journal of Public Policy*, vol.6, Part 2, pp.181-210.

O'Toole Jr., L. (1988), 'Strategies for Intergovernmental Management: Implementing Programs in Interorganizational Networks', *International Journal of Public Administration*, vol.11, pp.417-441.

O'Toole Jr., L. and Montjoy, R. S. (1984), 'International Policy Implementation: A Theoretical Perspective', *Public Administration Review*, Nov/Dec, pp.491-

503.

Offe, C. (1984), *Contradictions of the Welfare State*, Hutchinson, London.

Official Journal (1973), 'Environmental Action Programme. Resolution of the Council of the European Communities and of the Representatives of the Governments of the Member States, meeting within the Council of 22 November 1973 on the Programme of Action of the European Communities on the Environment', No C 112\1.

Official Journal (1977), 'Environmental Action Programme. Resolution of the Council of the European Communities and of the Representatives of the Governments of the Member States, meeting within the Council of 17 May 1977 on the Continuation and Implementation of a EC Policy and Action Programme on the Environment', No C 139\1.

Official Journal (1983), 'Environmental Action Programme. Resolution of the Council of the European Communities and of the Representatives of the Governments of the Member States, meeting within the Council of 7 February 1983 on the Continuation and Implementation of a EC Policy and Action Programme on the Environment', No C 46\1.

Official Journal (1985), 'Council Directive of 27 June 1985 on the assessment of the effects of certain public and private projects on the environment (85\337\EEC)', No L 175\40.

Official Journal (1987), 'Environmental Action Programme. Resolution of the Council of the European Communities and of the Representatives of the Governments of the Member States, meeting within the Council of 19 October 1987 on the Continuation and Implementation of a EC Policy and Action Programme on the Environment', No C 328\1.

Official Journal (1993), 'Environmental Action Programme. Towards Sustainability. Resolution of the Council and the Representatives of the Governments of the Member States, meeting within the Council of 1 February 1993 on a Community Programme of Policy and Action in relation to the Environment and Sustainable Development', No C 138\1.

Official Journal (1997), 'Council Directive 97/11/EC amending Council Directive of 27 June 1985 on the assessment of the effects of certain public and private projects on the environment (85\337\EEC)', No L 73\5.

Pag, S. and Wessels, W. (1988), 'Federal Republic of Germany', in H. Siedentopf and J. Ziller (eds), *Making European Policies Work. The Implementation of Community Legislation in the Member States*, Sage, London, pp.165-229.

Pressman, J.L. and Wildavsky, A. (1974), *Implementation: How great expectations in Washington are dashed in Oakland*, University of California, Los Angeles.

Puchala, D. (1975), 'Domestic Politics and Regional Harmonisation in the European Communities', *World Politics*, vol.27, pp.496-520.

Raschke, J. (1988), *Soziale Bewegungen. Ein historisch-systematischer Grundriss*, Campus.

Regierung von Mittelfranken (1995), *Umweltschutz im Regierungsbezirk Mittelfranken*, October, pamphlet.

Regierung von Oberbayern (no date), *Informationen zu Naturschutz und Landschaftspflege*, pamphlet.

Regierung von Schwaben (1996), *Neubau der Ortsumfahrung Gundelfingen-Lauingen der Bundesstrasse 16. Planfeststellungsbeschluss vom 28. November 1996.*

Regierung von Schwaben (1996), *Schwabeninitiative Beschleunigung von Genehmigungsverfahren.*

Regierung von Unterfranken (1996), *Umweltschutz im Regierungsbezirk Unterfranken*, April, pamphlet.

Rhodes, R. A. W. (1988), *Beyond Westminster and Whitehall. The sub-central governments of Britain*, Unwin Hyman, London.

Rhodes, R. A. W. (1997), *Understanding Governance. Policy Networks, Governance, Reflexivity and Accountability*, Open University Press, Buckingham.

Rhodes, R. A. W. and Marsh, D. (ed) (1992), *Policy Networks in British Government*, Clarendon Press, Oxford.

Rhodes, R. A. W. and Marsh, D. (eds) (1992), *Implementing Thatcherite Policies. Audit of an Era*, Open University Press, Buckingham.

Richardson, J. (1996), 'Eroding EU Policies: Implementation Gaps, Cheating and Re-steering', in J. Richardson (ed), *European Union. Power and Policy-Making*, Routledge, London, pp.278-294.

Robinson, M. (1992), *The Greening of British Party Politics*, Manchester University Press, Manchester.

Rometsch, D. (1995), *The Federal Republic of Germany and the European Union. Patterns of Institutional and Administrative Interaction*, Discussion Paper in German Studies, University of Birmingham, December.

Roth, R. and Rucht, D. (eds) (1991), *Neue Soziale Bewegungen in der Bundesrepublik Deutschland*, Bundeszentrale für politische Bildung, Bonn.

Sabatier, P. (1986), 'Top-down and Bottom-up Approaches to Implementation Research: A Critical Analysis and Suggested Synthesis', *Journal of Public Policy*, vol.6, Part 2, pp.21-48.

Sabatier, P. (1998), 'The advocacy coalition framework: revisions and relevance for Europe', *Journal of European Public Policy*, No.5, March, pp.98-130.

Sabatier, P. and Mazmanian, D. (1980), 'The Implementation of Public Policy: A Framework of Analysis', *Policy Studies*, vol.8, pp.538-560.

Sabatier, P. and Mazmanian, D. (eds) (1981), *Effective Policy Implementation*, Lexington Books, Massachusetts.

Sbragia, A. (1996), 'Environmental Policy: The Push-Pull Policy-making', in H. Wallace and W. Wallace (eds), *Policy-Making in the European Union*, Oxford University Press, Oxford, pp.235-255.

Scharpf, F. (1994), *Community and Autonomy. Multi-level Policy-Making in the*

European Union, European University Institute, Florence, Working Paper RSC No.94/1.

Schumann, W. (1991), 'EG-Forschung und Policy-Analyse. Zur Notwendigkeit den ganzen Elefanten zu erfassen.', *Politische Vierteljahresschrift*, 32. Jahrgang, Heft 2, pp.232-257.

Schwab, J. (1997), 'Die Umweltverträglichkeitsprüfung in der behördlichen Praxis', *Neue Verwaltungsrecht Zeitung*, Heft 5, pp.428-435.

Scott, A. et al (1994), 'Subsidiarity: A 'Europe of the Regions' versus the British Constitution?', *Journal of Common Market Studies*, vol.32, No.1, pp.47-67.

Scottish Executive (1999), *Scottish Circular* 15/99.

Scottish Executive (1999a), *Planning Advice Note* (PAN) 58.

Scottish Executive (1999b), *The Planning Bulletin: December 1999*.

Scottish Executive (2000), press release (SE2846), (3. November).

Scottish Natural Heritage (SNH) (1994), *Third Operational Plan - Review of 1993-94*.

Scottish Natural Heritage (SNH) (1995), *Work Programme 1994-95*.

Scottish Natural Heritage (SNH) (1997), *Corporate Plan 1994-95/ 1996-97*.

Scottish Office (1988), *Environmental Assessment (Scotland) Regulations 1988*, SI 1988 No.1221.

Scottish Office (1988), *Scottish Development Department Circular 13\1988*.

Scottish Office (1990), *Environmental Assessment - A Guide*, pamphlet, 6/90.

Scottish Office (1994*), Scottish Development Department Circular 26\1994*.

Scottish Office (1996), *Common Sense, Common Purpose. Sustainable Development in Scotland*, pamphlet.

Scottish Office (1997), *Scotland's Parliament*, White Paper presented to Parliament by the Secretary of State for Scotland by Command of Her Majesty, July.

Scottish Office and CoSLA (1993), *Local Environment Charter for Scotland*, September.

Scottish Parliament (2001), *Transport and Environment Committee publishes GMU Report*, press release (CTRAN01/2001), (23. January).

Scottish Wildlife and Countryside Link (1997), *People and the Environment: A Common Cause*, Conference Summary, Perth, 14 February.

Secretaries of State for Environment, Trade and Industry, Health, Education and Science, Scotland, Transport, Energy and Northern Ireland, the Minister of Agriculture, Fisheries and Food and the Secretaries of State for Employment and Wales (1990), *This Common Inheritance. Britain's Environmental Strategy*, White Paper presented to Parliament by Command of Her Majesty, September.

Seidel, R (no date), *UVP bei Industriestandorten am Beispiel eines Kraftwerk-Genehmigungsverfahrens nach BimschG*, Ansbach, Mittelfranken.

Siedentopf, H. (1990), *Die Umsetzung des Gemeinschaftsrechts durch die Verwaltungen der Mitgliedsstaaten*, Europa-Institut, Universität des

Saarlandes, Saarbrücken.

Siedentopf, H. and Ziller, J. (eds) (1988), *Making European Policies work. The Implementation of Community Legislation in the Member States*, Sage, London.

Sloat, A. (2001), *Scotland in the European Union: Expectations of the Scottish Parliament's Architects, Builders, and Tenants*, Scotland Europa Paper 22, (March).

Smith, J. A. (1990), *A Critical Appraisal of the Performance of the Environmental Assessment (Scotland) Regulations since their Introduction*, MSc Dissertation, University of Stirling, Stirling, September.

Smith, M. (1997), 'Brussels in environmental clampdown', *Financial Times*, 18/19. October, p.2.

Stadt Abensberg (no date), *Bürgerinformation Landschaftsplan der Stadt Abensberg*, pamphlet.

Strathclyde Regional Council (1994), *Environmental Action in Strathclyde. Charter for the Environment Strathclyde*, Chief Executive Department, pamphlet.

The Economist (1996), 'A Survey of Energy', supplement, 18. June.

The Economist (1997), 'Road Protests. Victory', 25. January, pp.31/32.

The Economist (1998), 'Environment and transport: Green Gordon', 21. March, p.41.

The Economist (1998), 'An invaluable environment', 18. April, p.105.

The Economist (1998), 'Train, planes, but not automobiles', 6. June, p.34.

The Guardian (2001), 'Landfill tax up 10% to deter dumping', 7. March, website.

The Herald (1992), 'Scotland Europa: Opening Night for Scotand's Voice in Europe', 27. May.

The Herald (1997), 'Scottish Office and Cosla in new accord', 22. August, p.7.

The Independent (1997), 'Brown puts focus on pollution and energy', 26. November, p.18.

The Scotsman (2001), 'Scotland could go it alone by 2013', website.

The Scotsman (2001), 'Labour gets new poll warning', website.

Treweek, J. R., Thompson, S., Veitch, N and Japp, C. (1993), 'Ecological Assessment of Proposed Road Developments: A Review of Environmental Statements', *Journal of Environmental Planning and Management*, vol.36, No.3, pp.295-307.

Vedder, E. (1990), 'Der aktuelle Stand der UVP-Gesetzgebung in der Bundesrepublik Deutschland und Bayern', *Inhalte und Umsetzung der Umweltverträglichkeitsprüfung (UVP)*, Laufener Seminarbeiträge 6/90, Akademie für Naturschutz und Landschaftspflege, pp.32-35.

Viebrock, J. (1992), 'Beschränkungen der UVP in der Verkehrswege-planungsbeschleunigung', *Neue Zeitschrift für Verwaltungsrecht*, 11.Jahrgang, pp.939-942.

Ward, A. (1993), 'The Right to an effective Remedy in European Community Law and Environmental Protection: A Case Study of UK Judicial Decisions concerning the Environmental Assessment Directive', *Journal of Environmental Law*, vol.5, Part 2, pp.221-244.

Ward, N., Buller, H. and Lowe, P. (1995), *Implementing European Environmental Policy at the Local Level: The British Experience with Water Quality Directives*, University of Newcastle upon Tyne, March.

Weale, A. et al (1991*), Controlling Pollution in the Round. Change and Choice in Environmental Legislation in Britain and Germany*, Anglo-German Foundation Project, London.

Weale, A. et al (1996), 'Environmental Administration in six European States: Secular Convergence or National Distinctiveness?', *Public Administration*, vol.74, Summer, pp.255-274.

Weale, A. (1996), 'Environmental rules and rule-making in the European Union', *Journal of European Public Policy*, vol.3, No.4, pp.594-611.

Weber, A. and Hellmann, U. (1990), 'Das Gesetz über die Umweltverträglichkeitsprüfung (UVP-Gesetz)', *Neue Juristische Wochenschrift*, Heft 27, pp.1625-1633.

Weber, J. (1993), *Environmental Planning as a Part of Urban Planning in the Federal Republic of Germany - The City of Munich as an Example*, Paper, Universidade Nova de Lisboa.

Webster, R. (1998), 'Environmental Collective Action. Stable patterns of co-operation and issue alliances at the European level', in J. Greenwood, and M. Aspinwall (ed), *Collective Action in the European Union. Interests and the new politics of association*, Routledge, London, pp.176-195.

Wegner, H. A. (1993), *Die Umweltpolitik der EG im Spannungsfeld zwischen Harmonisierungszwang und Subsidiaritätsprinzip*, Sonderdruck aus Berichte der Bayerischen Akademie für Naturschutz und Landschaftspflege, Nr.17.

Weidner, H. (1989), 'Die Umweltpolitik der konservativ-liberalen Regierung. Eine vorläufige Bilanz', *Aus Politik und Zeitgeschichte*, 47-48/89, 17. November, pp.16-28.

Werner, J. (1996), 'Das EU-Netzwerk für die Umwetzung und Vollzug des Umweltrechts', in G. Lübbe-Wolf (ed), *Der Vollzug des Europäischen Umweltrechts*, Erich Schmidt Verlag, Berlin, pp.131-138.

Williams, R. (1991), 'Direct Effect of EC Directive on Local Authorities', *The Cambridge Law Journal*, vol.50, Part 3, pp.382-384.

Wissmann, M. (1994), 'German Transport Policy after Unification', *Transport*, vol.28A, No.6, pp.453-458.

Wood, C. (1995), *Environmental Impact Assessment: A Comparative Review*, Longman, Essex.

Wood, C. and Jones, C. (1991), *Monitoring Environmental Assessment and Planning*, Department of the Environment Planning Research Programme.

Wood, C. and Jones, C. (1997), 'The Effect of Environmental Assessment on UK

Local Planning Authority Decisions', *Urban Studies*, vol.34, No.8, pp.1237-1257.

World Commission on Environment and Development (1987), *Our Common Future*, Brundtland Report, Oxford University Press, Oxford.

Young, S. C. (1993), 'Environmental Politics and the European Community', *Politics Review*, vol.2(3), February, pp.6-8.

Young, S. C. (1993), *The Politics of the Environment*, Baseline Books, Manchester.

Further information on the field research (including details on interviewees' written and verbal correspondence) is available from the author (a.c.brown@abdn.ac.uk).